The SAM-e Solution

THE
SAM-e
SOLUTION

Deborah Mitchell

Foreword by Steven J. Bock, M.D.

A LYNN SONBERG BOOK

WARNER BOOKS

A Time Warner Company

Warner Books, Inc., 1271 Avenue of the Americas, New York, NY 10020
Visit our Web site at www.twbookmark.com

 A Time Warner Company

Printed in the United States of America

First Warner Books Printing: October 1999

10 9 8 7 6 5 4 3 2 1

ISBN: 0-446-67637-3

LCCN: 99-65497

Cover design by Don Puckey

Text design by Stanley S. Drate/Folio Graphics Co. Inc.

Contents

Foreword

As a family physician and director of Rhinebeck Health Center, I have been treating depression and osteoarthritis for many years. I incorporate reliable complementary medical remedies into my practice whenever possible, so it has been with great interest that I have followed the research and development of SAM-e over the years, noting the promising results and safety reports my colleagues overseas have been documenting. But my interest was from afar, because SAM-e was not available in the United States. Some U.S. physicians and patients, however, were importing the supplement and were willing to accept the uncertainty of delivery for one reason: It worked for them. So by the time SAM-e hit the U.S. market, I was armed with what I believed to be sufficient clinical evidence and anecdotal reports to begin prescribing SAM-e to my patients.

Like so many of my colleagues, I know that conventional medicine does not always have the answers to patients' problems or questions. We know that people are tired of taking medications that often cause side effects that are worse than the condition they're treating. And we know that our patients are curious about and are trying alternative remedies. The arrival of SAM-e has opened a new, exciting door of opportunity, both for us as physicians and for the patients we treat: the chance to offer relief from the darkness of depression or from the pain and limitations of osteoarthritis from a natural product that is virtually without side effects.

I sat down with each patient who was a likely candidate for SAM-e and we discussed the idea. I explained what the clinical studies showed, how Europeans had been using SAM-e for decades, and that the possibility of side effects was very low. All of the patients I spoke with were enthused about using SAM-e, and so my own personal study began. The bulk of the new nutritional supplements that are introduced to the marketplace are probably somewhat helpful for a large percentage of people. Indeed, many people derive benefits from St.-John's-wort, for example, for the treatment of depression. But I find that SAM-e works faster than St.-John's-wort, faster than conventional antidepressants, is just as effective, and doesn't have the disadvantage of side effects. My patients with osteoarthritis are experiencing less pain and greater mobility and are confident that SAM-e is helping rebuild their damaged cartilage. I share their enthusiasm and hope, and so far I see no reason to believe SAM-e is not delivering what it has promised for treatment of depression and osteoarthritis.

The "poster child" patient for SAM-e in my practice is "Sylvia," a woman I have been treating for depression, osteoarthritis, and liver problems. Our plan was to treat her with 1,200 mg daily of SAM-e for her osteoarthritis. After only three days, she came to me and said, with total amazement and joy, that her depression had lifted dramatically, something the Prozac and lithium she had been taking had been unable to do. Needless to say, she is continuing with SAM-e and is hopeful she will soon experience relief from her osteoarthritis symptoms.

In a sense our work here in the United States is just beginning. The conscientious efforts of our overseas colleagues need to be recognized and applauded but, most important, continued. SAM-e is prescribed in fourteen countries for one or more conditions. People in these countries have been getting relief from depression, osteoarthritis, liver disease, and even fibromyalgia. So this is where the challenge begins: the challenge to conventional health-care practitioners and consumers alike. Both need to react responsibly to the challenge. If physicians and patients ignore the glitzy rhetoric and instead listen to the researchers and read the studies, what they will find is a credible argument on the side of SAM-e for the treatment of depression, osteoarthritis, and liver disease, as well as some very exciting possibilities in other areas. Physicians have an opportunity to learn along with their patients. They also need to demand quality products from manufacturers and to be willing to evaluate the past and current studies with an open mind.

If you, as a consumer and as an individual with health problems, want to have some control over your health, then you need to educate yourself about SAM-e. This book helps you do that. It tells you what SAM-e is, how it works in the body, and, more important, it explains the studies that have led researchers and physicians to make the claims they do about SAM-e.

If you want to be told that SAM-e is a miracle cure, you've come to the wrong book. *The SAM-e Solution* explains the studies that have been done that support or suggest the use of SAM-e for treatment of depression, osteoarthritis, liver disorders, fibromyalgia, migraine, Alzheimer's disease, Parkinson's disease, and several other conditions. But it also tells you that not all the studies have been positive and that on rare occasions there may be some mild, temporary side effects. The book also explains how to use SAM-e, either alone or to enhance the benefits of another remedy, whether it be conventional or alternative; what to look for when buying SAM-e to ensure you get a viable, potent product; and answers the questions that come to mind when considering a new supplement, including issues of safety, dosing, effectiveness, and other topics. Most of the supplements that come to market do not have staying power; I believe SAM-e will be an exception. If the study results continue to be positive and the adverse effects remain negligible, SAM-e holds great promise for people with depression, osteoarthritis, liver disease, and perhaps even several more serious health problems. *The SAM-e Solution* introduces you to the future and challenges you to be part of it.

—STEVEN J. BOCK, M.D.

Introduction

Do you have a chronic case of the blues?

Are you one of approximately 17 million Americans who have been diagnosed with clinical depression?

Is fibromyalgia preventing you from doing all the things you want to do?

Do you wake up each day stiff and in pain from osteoarthritis, knowing that the only relief you get every day is from drugs that are ultimately causing you more harm than good?

Are you and your liver battling the destructive power of cirrhosis or hepatitis?

Are you or someone you love experiencing short-term memory loss, loss of concentration, or other mental lapses usually associated with aging?

Would you like to find a way to fight the damaging effects of aging?

If you answered yes to one or more of these questions, then you need to meet SAM-e (pronounced "Sammy"). SAM-e, or in scientific terms S-adenosyl-L-methionine, is a new dietary supplement with a friendly name, a long history, and a promising future. Although new to the United States, SAM-e is no stranger to our friends overseas. It has been used widely throughout Europe for more than twenty years as a treatment for depression and osteoarthritis, and more recently for fibromyalgia and liver disorders, such as cirrhosis and hepatitis. Researchers are finding other possible uses for SAM-e as they continue to investigate its benefits for people with heart disease, and for those with conditions involving learning and memory problems associated with aging, such as Alzheimer's disease. Many other medical conditions have also been the focus of preliminary studies. SAM-e may hold significant promise for conditions including migraine, attention deficit hyperactivity disorder, and Parkinson's disease.

SAM-e has been available in Europe since 1975. It has gained widespread acceptance by the medical community there and become a prescription medication in Germany, Italy, Russia, and Spain. In Germany, SAM-e is an approved treatment for arthritis. In Italy, it outsells the antidepressant Prozac, a drug Americans have had a love affair with for more than a decade—prescription sales for Prozac top $2 billion a year, and more than 12 million people around the world (including more than 6 million Americans) take it.

SAM-e Meets the United States

In February 1999, SAM-e made its debut in the United States as a nonprescription, over-the-counter food supplement, available in local neighborhood pharmacies, nutrition stores, grocery stores, and large chain department stores, as well as through mail order and the Internet. Unlike some supplements that seem to "burst" upon the scene with little else but a long trail of anecdotal reports to support their claims, SAM-e hit U.S. shelves with an arsenal of scientific studies supporting the claims being made about the supplement. Not that anecdotal reports are bad; in fact they are very helpful in advancing and bringing a new supplement to market and encouraging scientific research. But most people feel more secure knowing there are scientific studies to support the claims for the supplement they are considering. They feel more confident knowing that millions of people have already taken the supplement and have had good results. At the same time, the majority of people do not have the time to go through hundreds of studies and thousands of pages of evidence to find out what SAM-e is all about.

The SAM-e Solution has done the work for you. Within these pages are the results, findings, and conclusions from more than a quarter century worth of studies that chronicle SAM-e's story: from the time it was formulated in Italy in 1973, when researchers there first discovered its potential as a treatment for depression, to today, when researchers have identified its value in the treatment of other conditions that take an enormous toll on the health of men, women, and children everywhere. This book provides you with information you can use to improve the quality of your life or the life of someone you care about. It gives you the information you need to make an informed choice when purchasing SAM-e; information you can take with you to your health-care practitioner so the two of you can talk about your treatment plans. If at any time you want to learn more about any of the studies or findings you read about in a particular chapter, *The SAM-e Solution* has placed the relevant information in notes at the end of each chapter.

In Chapter 1, you will learn what SAM-e is, where it comes from, and why so many people are excited about it. Once you read about all the things SAM-e can do, you'll wonder how it can do so much. In Chapter 2 you will read about the secrets of SAM-e and how it works in the body. Chapters 3 and 4 take a close look at the condition SAM-e is most often associated with—depression. You will read about the different types of depression, its symptoms, the different pharmacological treatments that are available and the problems with them, how SAM-e stacks up against other treatments for depression, both conventional and alternative, and the studies to back up the claims.

Chapters 5 through 7 explore other conditions researchers have found that respond best to SAM-e. In Chapter 5 you will read how SAM-e can work to restore damaged and lost cartilage while it relieves the often debilitating symptoms of osteoarthritis. Chapter 6 will be welcome news to the more than 10 million people, mostly women, who suffer with the pain and depression associated with fibromyalgia. Individuals who live with hepatitis, cirrhosis, or other liver disorders can benefit from the information and research findings linking SAM-e and liver health covered in Chapter 7.

In Chapter 8, you will learn about the other conditions researchers have been able to link with SAM-e and its power to heal, including heart disease, Alzheimer's disease, Parkinson's disease, migraine, attention deficit disorder, and cognitive and mental conditions associated with the aging brain. Chapter 9 gives you the basics on which forms of SAM-e are the most effective, how to take SAM-e and when, and who can take it. It also answers questions about safety and how to use SAM-e with other supplements or drugs. The final chapter discusses the miscellaneous questions that don't fall conveniently into any other chapter but which deserve to be answered, and brings together some of the most often asked questions in a convenient place so you don't have to search through the book. Finally, the Epilogue looks forward to the future of SAM-e as the momentum of research continues and investigators continue to add to the solid base of knowledge they have accumulated about the benefits of SAM-e and how it can improve the quality of life.

As you prepare to turn the page to read those that follow, bear in mind that although a lot of research has been conducted on SAM-e overseas, much remains to be done in the United States. New studies will greatly expand our knowledge about how SAM-e works and the conditions for which it is best suited. Chapter 1 starts you on your way to meet SAM-e.

1

Meet SAM-e

When a new supplement enters the marketplace and studies show it to be beneficial for the treatment of any of the "big" or common medical conditions, such as depression, osteoarthritis, osteoporosis, and aging, the public is suddenly inundated with claims about the product. It often becomes difficult to separate the hype from the truth, and when people are in pain or have waited a long time for relief from their illness or symptoms, every word they hear or read rings true, or so it seems. Time often proves many of these supplements to be less than originally advertised, yet a great number of people do derive some benefits from them.

SAM-e is the latest entry into the supplement arena. While it comes with fanfare and claims, it also comes with what some experts say is solid proof that it is better than other treatments for depression, osteoarthritis, liver disease, and several other conditions, proof gathered over twenty years. It comes with a substantial history of use among millions of Europeans during the last twenty-plus years. Some researchers and physicians are saying that although it is not a miracle cure, it certainly has some outstanding abilities that can surpass those of drugs and other remedies currently being used to treat these conditions. To help you separate the hype from the truth, the claims from the scientific findings, you need to start at the beginning and discover what this new substance is and how it works.

What Is SAM-e?

SAM-e is a compound that occurs naturally in the body and is in every one of your cells. It is actually the result of a union between two substances found in the body: the amino acid L-methionine and ATP (adenosine triphosphate). L-methionine (usually written as methionine) is an essential amino acid, which means the body needs it to perform vital functions, yet the body is unable to synthesize it. Therefore it must be obtained from the diet. ATP is a compound composed of adenosine and three acid groups. As the body's primary energy molecule, it is involved in nearly every process in the body. Like methionine, it is also found in all cells, but it is particularly prevalent in muscle cells. When ATP is split by the action of enzymes, energy is produced. When you combine these two powerhouses—methionine and ATP—you get the dynamic compound that is SAM-e.

And dynamic it is. SAM-e is involved in approximately forty different biochemical reactions in the body. Here are just a few of the processes in which SAM-e plays a crucial role:

- It aids the synthesis of neurotransmitters in the brain, including serotonin and dopamine, which results in an enhanced mood.

- It facilitates the chemical processes involved in the control of the inflammation and pain of osteoarthritis and other inflammatory conditions, which results in pain relief.
- It promotes the production of enzyme antioxidants and several essential biochemical reactions such as detoxification of cells, which helps rid the body of various disease-causing substances.
- It supports vital functions that maintain a healthy liver.
- It protects the body against the harmful impact of homocysteine, a substance that has a negative impact on the heart and cardiovascular health.
- It is necessary for the synthesis of melatonin, a hormone that helps regulate the awake-sleep cycle.

SAM-e shares a characteristic with many other substances found in the body, including vitamins, minerals, hormones, and enzymes: its level in the body declines with age. Thus virtually everyone can expect to experience some impact from declining levels of SAM-e, even if it's only memory loss with advanced age. Some scientists believe the way to manage that decline may be to supplement with SAM-e.

However, older people are not the only ones who can have low amounts of SAM-e in their body. During the course of their research, investigators have found that low SAM-e levels are characteristic of many different medical conditions, which are discussed at length in this book. They also found that levels of SAM-e are not the same for men and women, or even for children and adults. Men have slightly higher blood levels of SAM-e than women, and children have about seven times greater levels than adults. These findings have been part of the fuel that has sparked the large number of research studies involving SAM-e over the last twenty-five years.

Where Does SAM-e Come From?

When people learn that low levels of SAM-e are associated with many different medical conditions as well as aging, they typically want to make sure they get enough SAM-e to help prevent these problems. Before you learn how and where to get enough SAM-e, you need to know where SAM-e comes from.

SAM-e can come from two natural sources: one direct, and one indirect (actually three if you count SAM-e supplements, but for this discussion let's look at the dietary sources).

A small amount of SAM-e is available directly from food, yet this source is not reliable. Unlike the stable form of SAM-e found in properly processed supplements, the type available in living organisms is highly unstable.

The cells in our bodies are also able to make a small amount of methionine, which is used to manufacture SAM-e, but not enough to meet the body's high demands. Therefore, it is important to get an adequate amount of methionine through your diet. A food plan that consists of 20 to 30 percent quality protein foods, such as fish, soybeans, legumes, and whole grains, should provide you with an adequate amount of the amino acid methionine. The body manufactures SAM-e by combining this amino acid with the energy molecule ATP, which is also found naturally in the body.

But consuming adequate amounts of methionine is not always a guarantee that your body will produce enough SAM-e. Methionine production also requires the help of several other nutrients, especially vitamins B_6, B_{12}, folic acid, and choline (part of the B complex of vitamins). Without adequate levels of these vitamins, methionine production declines. There is also a strong correlation between a deficiency of vitamin B_6, B_{12}, and folic acid, and depression and psychiatric disorders, including dementia. (This relationship is discussed in more detail in Chapter 4.)

Several other B vitamins also play a role in depression. Vitamin B_1, for example, has been linked with depression and anxiety, while niacin (vitamin B_3) is essential for the proper function of the nervous system. The relationship between methionine and the B vitamins supports the intimate connection between methionine, SAM-e, and depression. (For information on taking supplements with SAM-e, including recommended dosages, see Chapter 9.)

But for many people, even maintaining a nutritious diet with adequate protein and B vitamins is not enough to keep methionine at a level that best promotes the manufacture of SAM-e. Studies show that many people with depression, fibromyalgia, osteoarthritis, Alzheimer's disease, heart disease, migraine, liver disorders, and other conditions have low levels of SAM-e. Also, since SAM-e levels decline with age, it seems we will all eventually need supplementation. So why can't you boost your SAM-e levels by eating a lot more protein or, better yet, by taking supplements of methionine so your body will produce more SAM-e?

It seems to make sense that you could raise methionine levels, and thus SAM-e levels, if you were to eat a very high-protein diet. However, not only won't eating large amounts of protein guarantee a rise in either of these substances, it also can have damaging effects on your body. High protein intake is associated with liver and kidney problems and is a major cause of osteoporosis. Even if you were to increase your intake of methionine, you would also need to increase your intake of B vitamins, especially B_6, B_{12}, folic acid, and choline. The same holds true if you were to take methionine supplements, in which case you would also need to take a supplement that contained all the amino acids, because increasing your intake of just one amino acid can disrupt the balance of amino acids in the body.

As you can see, forcing your body to create more SAM-e is not a simple matter. You can, however, take SAM-e as a supplement, a synthesized form available in a tablet that will benefit your body. Then, there are several vitamins and other supplements you can take that can enhance your body's ability to assimilate and utilize SAM-e. (These issues are discussed later, in Chapter 9.)

How Was SAM-e Discovered?

As often happens in science, serendipity was the trigger that launched a whole new direction to a line of research. While investigating the use of SAM-e in people with schizophrenia, Italian researchers noted that the patients in their study were getting less and less depressed. That unexpected discovery started a flurry of studies into its effect on depression. Eventually those studies led researchers beyond SAM-e's connection with depression to its role in liver function, heart disease, joint and muscle inflammation, brain function, cell toxicity, and cancer. It seemed that the more scientists looked, the more things SAM-e was involved in.

Although it's true that SAM-e has been enjoying widespread use in Europe since it was first cultured in an Italian laboratory in 1972, SAM-e actually got its beginnings in the United States in 1952 at the National Institutes of Health. After that, it seemed to vanish from the American scene until interest was sparked by European studies in the 1970s. Even then, American participation in studies was sparse until later years.

This is easy to understand if you consider how drugs are brought to market in the United States. Drug research in the United States is funded primarily by pharmaceutical companies. It costs an enormous amount of money and takes such a long time (usually a decade or more) to conduct all the necessary research and testing required by the Food and Drug Administration (FDA) before a new drug can be patented and marketed to the American public. SAM-e is a natural substance produced by the body, it is not a drug. In the United States, it has traditionally been very difficult to patent natural substances. Without a patent and the potential to make a significant profit, pharmaceutical companies were not often motivated to invest their funds in natural products like SAM-e. So while people in European nations have long reaped the benefits of many natural remedies, individuals in the United States have had limited or no access to them.

All of that changed in 1994 when Congress passed the Dietary Supplement Health and Education Act (DSHEA). This law allows manufacturers to market any supplement that demonstrates a good safety record. The DSHEA has opened the door to many supplements and natural remedies so they can be sold on U.S. shelves without FDA approval. SAM-e is one of those natural supplements. However, manufacturers are not allowed to make any medical claims about SAM-e or any other dietary supplements and can only hint at their benefits with words such as "improves mood" and "promotes joint health" rather than "treats depression and mood disorders" and "treats arthritis."

One unfortunate aspect of the DSHEA is that it does not impose strict quality standards on dietary supplements, as the Food and Drug Administration (FDA) does for prescription drugs. Because supplements are considered to be "food," manufacturers are not required to test the products and report the results. Often, consumers are buying completely ineffective supplements that contain little or none of the ingredients they believe they are getting. That is why consumers are urged to buy their supplements from reputable manufacturers only. In some cases, it is even recommended that you get your supplement tested at an independent laboratory to check its potency. (Some retailers will do this as a service for you.) In the case of SAM-e, potency is a very critical issue and depends on how the supplement is produced and packaged, as you will learn. (This topic is discussed in more detail in Chapter 9.)

So, forty-seven years after it was first discovered in the United States, SAM-e has returned home. And it's no secret that it can be an effective treatment for many different ailments.

Why Are People Talking about SAM-e?

Would you be excited if:

- You could take a natural antidepressant that has little or no side effects and that works in half the time—sometimes within days—of other, conventional antidepressants?
- You could get relief from the soreness and depression that are the hallmarks of fibromyalgia without worrying about side effects or possible drug interactions from other medications you may be taking?
- You could treat your osteoarthritis with a safe, natural substance that not only relieves the pain, inflammation, and stiffness associated with the disease but may also trigger regeneration of the cartilage and lubricate the joints— all without the gastrointestinal symptoms associated with nonsteroidal anti-inflammatory drugs?
- You could reverse and possibly prevent aging of your brain and the symptoms that go along with it, such as short-term memory loss and difficulty concentrating?

The SAM-e Solution wants to share the excitement with you by letting you in on the SAM-e story. Let's begin by getting a better understanding of the mechanisms that make it possible for SAM-e to do what it does: "The Secrets of SAM-e."

NOTES

Carney, M. W. P., et al. "Red cell folate concentrations in psychiatric patients." *J Affective Disorders* 19 (1990): 207–213.

Cooper, C. "Dietary protein intake and bone mass in women." *Calcif Tissue Int* 58 (1996): 320–325.

Crellin, R., Bottiglieri, T., and Reynolds, E. H. "Folates and psychiatric disorders: Clinical potential." *Drugs* 45 (1993): 623–636.

Godfrey, P. S. A., et al. "Enhancement of recovery from psychiatric illness by methyl folate." *Lancet* 336 (1990): 392–395.

Licata, A., et al. "Acute effects of dietary protein on calcium metabolism in patients with osteoporosis." *J Geron* 36 (1981): 14–19.

Murray, M., *Encyclopedia of Natural Medicine.* Rocklin, CA: Prima Publishing, 1998.

Newsweek, "The 'Sammy' Solution," 22 March 1999.

Reynolds, E., et al. "Folate deficiency in depressive illness." *Br J Psychiat* 117 (1970): 287–292.

Zechner, O., et al. "Nutritional risk factors in urinary stone disease." *J Urol* 125 (1981): 51–55.

2

The Secrets of SAM-e

When you're in pain and you're about to take an over-the-counter or prescription medication, do you ever ask yourself, "How does this pill know where I hurt? How does this medicine work?" Is it magic? The answer is no, but it may seem like it is, especially if the medication you take or give to someone else provides relief you're looking for and need.

SAM-e is a powerful supplement because it can work so quickly for people who are suffering with depression; because it is involved in at least forty different processes in the body; because it can provide dramatic relief for some of the most prevalent and troublesome conditions that affect people today. Some people claim no other nondrug antidepressant on the market has been the subject of the thorough testing and research that has gone into bringing SAM-e to you, although St.-John's-wort certainly rivals SAM-e for that honor. Dozens of studies have been published that document SAM-e's effectiveness and benefits in the treatment of depression. Along the way, and often serendipitiously, researchers have found several other conditions that respond well to SAM-e. Throughout the years of research, many questions have arisen, some of which can be answered better than others. Questions such as: How does SAM-e work? What are the mechanisms that make it so effective in so many different processes? This chapter answers these questions for you the best it can, with the information that has been gathered thus far.

Uncovering SAM-e's Secrets

Chapter 1 noted that SAM-e is the result of a union between the essential amino acid methionine, which is produced in the body, and ATP, the fuel cells produced to provide your body with energy. This union is made possible because an enzyme called methionine S-adenosyltransferase (MAT) acts as a matchmaker and catalyzes (or stimulates) the reaction between methionine and ATP. SAM-e is involved in several key processes which in turn are responsible for dozens of complex processes that keep the body functioning. One of the key processes is called methylation, which is actually an umbrella term for many types of chemical reactions. Several other processes are also discussed in this chapter, including transsulfuration, the production of polyamines, and the critical role SAM-e plays in the conversion of serotonin into melatonin. All of these processes make it possible for SAM-e to provide its wide-ranging benefits for the body—from relieving depression to eliminating toxins and helping rebuild cartilage in the joints.

Methylation

Methylation is a chemical reaction in which a methyl group—composed of one carbon and three hydrogen atoms—is passed from one molecule to another. This simple transfer of a methyl group can activate at least forty, and some say as many as one hundred, different processes in the body, such as the synthesis of DNA, proteins, and neurotransmitters, and processes that help suppress viruses, promote bone density, and protect against heart disease. SAM-e is a methyl donor; that is, it gives up its methyl groups to molecules that are involved in different biochemical reactions. SAM-e's "donations" transform the receiving molecules into active substances that are involved in various vital activities that are responsible for health and well-being. For example, "Methylation is implicated in the etiology [cause] of psychiatric illness," says Dr. B. L. Kagan in an article in the *American Journal of Psychiatry*, and particularly in depression.

Methylation of Proteins and Its Role in Depression

Proteins are organic compounds composed of chains of amino acids. They are the foundation for many of the body's components, including organs, tissues, muscles, and hormones. When SAM-e, as a methyl donor, gives methyl groups to protein molecules, this transfer can activate receptor sites on the receiving molecules, which in turn increases both the production of neurotransmitters (especially those involved in depression—serotonin, dopamine, and norepinephrine) and the number of receptors on the cells.

Both of these reactions are key in the prevention of depression: adequate levels of neurotransmitters are necessary to prevent the processes that lead to depression, and sufficient receptor sites are needed to ensure those processes will continue. The role of neurotransmitters in depression is explained further in Chapter 3.

Methylation of Phospholipids and Its Role in Aging

SAM-e also donates methyl groups to phospholipids. Phospholipids are fatlike substances that are found in all cell membranes. The transfer of the methyl group from SAM-e to phospholipids results in the production of phosphatidylcholine, also known as lecithin, which resides in cell membranes, as well as other places in the body. If you think of a cell membrane as a border between two countries—the outside of the cell being one country and the inside of the cell being another—then the lecithin molecules are the guards at the gates along the border. One job of lecithin is to regulate what goes in and out of the cell: nutrients and other beneficial substances go in; toxins and waste materials go out.

Another phospholipid found in cell membranes is cholesterol. The lecithin SAM-e creates helps keep the cell membranes pliable and permeable, and prevents the cholesterol from making them hard. When we are young, there is usually no problem, because SAM-e and lecithin levels are generally adequate, cholesterol levels are lower, and substances can move freely and properly through the border. But as we grow older, SAM-e levels decline, and so do lecithin levels. This process allows the ratio of lecithin-to-cholesterol to balance in favor of cholesterol, which makes the cell membranes stiff and less permeable. Pliable membranes have more receptors and so are able to transmit signals better.

Thus, the progression is this: the decline of SAM-e levels as we age, plus the decline of methylation as we age, leads to a decline in lecithin in cell membranes, which leads to a predominance of cholesterol in cell membranes, which contributes to aging and disease. By maintaining levels of SAM-e as we get older, we can help prevent diseases associated with aging, such as heart disease and dementia. But aging is not the only reason SAM-e levels can be low in the body. People with depression, osteoarthritis, fibromyalgia, liver disorders, heart disease, migraine, Alzheimer's disease, Parkinson's disease, and other conditions also have been found to have low levels of SAM-e. The potential for hardening of the cell membranes and a loss of cell integrity can occur at nearly any age. Supplemental SAM-e can help the body battle these diseases and ailments.

Methylation of DNA and Its Role in Cell Health

SAM-e also participates in a critical methylation process that involves DNA. Deoxyribonucleic acid, or DNA, is found in all cells and is the bearer of the genetic information for all living things. When SAM-e donates a methyl group to DNA, it causes the genes to become either active or inactive. Activated genes perform processes that result in cell growth, reproduction, and repair. In this way, supplemental SAM-e may promote the manufacture of cartilage in people with osteoarthritis. Activated genes also regulate production of immune system cells that help fight infection and disease, promote wound healing, and can stop the growth of tumors. Inactivated genes also can halt tumor growth. A decline in SAM-e levels, therefore, may result in poorer cell growth and repair, as occurs in aging, as well as a less effective immune system response to infection and cancer.

SAM-e, Serotonin, and Melatonin

The methylation process that occurs between SAM-e and serotonin to produce melatonin is an especially interesting one because this relationship is involved with your body's internal clock and your sleep and wake patterns.

Melatonin is a hormone that is secreted by the pineal gland, which is in the base of the brain. You've probably heard about melatonin as a supplement you can take to ward off jet lag or to treat insomnia. It is used for these conditions because one of the many functions of melatonin is the synchronization of hormone release, often called the *circadian rhythm*. The circadian rhythm is the body's internal clock, which controls the release of certain hormones at certain times of the day and night and thus controls sleep and awake periods. Release of melatonin is triggered by darkness and suppressed by light. A deficiency of melatonin or a disruption in its production can cause insomnia and other sleep problems.

SAM-e is essential for the manufacture of melatonin from serotonin because it donates a methyl group molecule to the enzyme that converts a form of serotonin into melatonin. The SAM-e and serotonin produced during the day assist in the manufacture of melatonin at night. Actually, SAM-e and melatonin have a seesaw relationship: when melatonin levels rise at night, SAM-e levels drop; when the sun comes up, melatonin levels drop and SAM-e's rise.

Scientists have tracked the rise and fall of SAM-e and melatonin in the body according to the time of day and night. For example, about thirty minutes before the sun goes down, levels of SAM-e peak. They stay high for about one hour, then drop dramatically. As soon as SAM-e levels fall, melatonin levels start to rise. They continue to climb for about four hours, and at about the fifth hour, melatonin peaks while SAM-e reaches its lowest point. Melatonin stays high until three hours before the sun comes up, when it falls suddenly. In the meantime, SAM-e levels are on the rise again. If there is an insufficient supply of SAM-e during the day, neither serotonin nor melatonin can be manufactured, and an imbalance in the amount of serotonin and melatonin can cause sleep problems.

Transsulfuration

This is a fancy word for a process that involves the transformation of SAM-e into sulfur-containing compounds. Sulfur is an essential mineral that is found in all cells in the body. Without sulfur life could not exist. SAM-e plays a key role in maintaining sulfur levels in the body and in promoting the processes in which this mineral is involved.

Transsulfuration begins when SAM-e donates its methyl group and is transformed into S-adenosylhomocysteine, or SAH. The SAH molecules then donate sulfur ("S") to other molecules, creating the sulfur-containing amino acids taurine and cysteine. Cysteine combines with two other amino acids, glycine and glutamic acid, and forms the amino acid glutathione. The SAH molecule then gives up the adenosine component that was gained during the union of methionine and ATP, leaving the amino acid homocysteine behind. (Homocysteine is important in heart disease; see Chapter 8.)

Although cysteine and taurine are important substances in the body, the focus here is on glutathione, a powerful antioxidant that cannot be manufactured without SAM-e, and homocysteine, a byproduct of transsulfuration.

Glutathione

Glutathione is known as a free-radical scavenger. Free radicals are highly charged molecules that have an odd number of electrons, so they seek out other molecules in an effort to pair up their electrons. The targets of their search are other unstable molecules, such as those in toxic substances (e.g., herbicides, cigarette smoke, gas fumes, food preservatives) or those associated with stress, chemotherapy, and radiation. If free radicals are allowed to run rampant and to keep pairing up, they will go on to produce more free radicals and cause damage throughout the body. The damage caused by free radicals is called *oxidation*. Premature aging is just one of the effects of free-radical damage. Heart disease, cancer, and more than sixty other diseases have been linked with oxidation.

The way to curtail or stop free-radical damage is with antioxidants. Glutathione is an especially powerful antioxidant. In fact, many experts call glutathione an age buster, because it is the primary compound responsible for neutralizing free radicals, which helps slow the aging process. Studies show that when glutathione-deficient cells are provided with this amino acid, they are able to regenerate and provide immune system protection.

The liver, where concentrations of glutathione and SAM-e are the highest in the body, is also where glutathione performs one of its most important functions: detoxification. The liver is involved in the breakdown and elimination of toxic and other waste materials that enter the body through food, water, drugs, or other means, or that are the byproducts of normal metabolic processes. The ability of the liver to perform detoxification depends largely on the level of glutathione in the liver. Glutathione seeks out toxins in the liver, attaches itself to them, and makes them water soluble so they can be eliminated in the urine. Liver damage occurs when glutathione levels are low, or when the liver is so overburdened with toxins, a situation that is characteristic of alcohol abuse or a viral infection, that the glutathione in the liver cannot rid the liver of the poisons. SAM-e has proved to be effective in the treatment of liver disease because it promotes production of glutathione and, because it is a methyl donor, it also provides other benefits to the liver, which are discussed in more detail in Chapter 7.

Although glutathione is prominent in the liver, it is also found in other parts of the body. High concentrations of glutathione are found in the lens of the eye, where it helps prevent the formation of cataracts. It also helps reduce inflammation in the joints, which makes it an effective treatment for osteoarthritis and bursitis. There is also some evidence that glutathione may help prevent viral replication in patients with the AIDS virus.

Why not just take a glutathione supplement for

liver problems? Isn't it easier and faster to bypass the whole transsulfuration process by taking glutathione from the start? Experts disagree about the value of glutathione supplements. Many say the amino acid is destroyed in the gastrointestinal tract and so does not reach the bloodstream. Supplementation with SAM-e avoids this problem. SAM-e has been shown to raise glutathione levels and to be an effective treatment for liver disorders (see Chapter 7).

Homocysteine

A byproduct of transsulfuration is homocysteine, an amino acid that is produced by all cells in the body. While the body can make good use of small amounts of homocysteine, high levels can cause significant problems, even death. Within the last few years, homocysteine has been identified as a promoter of atherosclerosis and as an important independent risk factor for heart attack, stroke, and peripheral vascular disease. Studies published in the *American Journal of Cardiology* and the *New England Journal of Medicine* show that approximately 25 to 40 percent of patients with heart disease have elevated levels of homocysteine. There have even been suggestions that increased concentrations of homocysteine in the blood in postmenopausal women may help promote osteoporosis by causing a breakdown in the bone matrix.

It seems clear that high levels of homocysteine are not desirable. Again, SAM-e may play a significant role, along with its helpers folic acid and vitamin B_{12}, as well as vitamin B_6, in keeping homocysteine levels within safe bounds. A deficiency of these three B vitamins is associated with elevated levels of homocysteine, and research has shown that taking supplements of all three can lower the level of the amino acid. Part of the process of reducing the amount of homocysteine in the body involves the conversion of the harmful amino acid into methionine, which in turn is used to make SAM-e. SAM-e then stimulates an enzyme to transform the homocysteine into glutathione.

This brief look at homocysteine and SAM-e suggests some benefits in the prevention or treatment of heart disease. This topic and the studies that have been done on it are covered in more detail in Chapter 8.

SAM-e and Polyamines

Another important function of SAM-e is its role in the production of polyamines. Polyamines are nitrogen-containing compounds that are involved in gene expression (how certain characteristics are expressed, such as eye color and skin tone) and the binding and repairing of DNA. SAM-e is the sole contributor of an aminopropyl group that links up with a polyamine called putrescine, which in turn forms two other polyamines, spermidine and then spermine. As their names suggest, spermidine and spermine are found in semen, and spermine is also found in other tissues in the body. Both of these polyamines are essential for cell differentiation and growth and protein synthesis. Spermidine may improve protein efficiency and decrease the effects of destructive forces on protein in the body. Spermidine and spermine both have some anti-inflammatory and painkilling properties and, like glutathione, are free-radical scavengers. Thus SAM-e's role in the production of polyamines may make it indirectly involved in reducing the pain and inflammation associated with diseases such as fibromyalgia and osteoarthritis.

Secrets at Work: Getting In

Now that you've read through this chapter, you can see that SAM-e, whether it is produced naturally by the body or taken in supplement form, has a multitude of functions to perform. It's as if there is a blueprint of action for each process, the parts of which are well orchestrated by SAM-e.

This overview of SAM-e's secrets reveals the nuts and bolts of its activities in the body. In the next six chapters you will read in detail about how SAM-e's great versatility allows it to have far-reaching effects in the body and to be effective in relieving various medical conditions, some of which appear to have little or no connection with each other: depression, arthritis, fibromyalgia, liver disorders, heart disease, Alzheimer's disease, Parkinson's disease, and others.

NOTES

Brattstrom, L. E., Hultberg, G. L., and Hardebo, J. E. "Folic acid responsive postmenopausal homocysteinemia." *Metabolism* 34 (1985): 1073–1077.

Chiang, P. K., et al. "S-Adenosylmethione and methylation." *FASEB J* 10(4) (10 March 1996): 471–480.

Clarke, R., et al. "Hyperhomocysteinemia: An independent risk factor for vascular disease." *New Engl J Med* 324 (1991): 1149–1155.

Glueck, C. J., et al. "Evidence that homocysteine is an independent risk factor for atherosclerosis in hyperlipidemic patients." *Am J Cardio* 75 (1995): 132–136.

Kagin, B. L., Sultzer, D. L., Rosenlicht, N., Gerner, R. H. "Oral s-adenosylmethionine in depression: A randomized double-blind placebo-controlled trial." *Am J Psychiatry* 147(5) (May 1990): 591–595.

Rathbun, W., and Hanson, S. "Glutathione metabolic pathway as a scavenging system in the lens." *Ophthal Res* 11 (1979): 172–176.

Schumacher, H. Ralph, Jr. "Osteoarthritis: The clinical picture, pathogenesis, and management with studies on a new therapeutic agent, S-adenosylmethionine."*American Journal of Medicine* 83 (Suppl 5A).

Staal, F. J., et al. "Glutathione deficiency and human immunodeficiency virus infection." *Lancet* 339 (1992): 909–912.

Ubbink, J. B., van der Merwe, W. I., and Delport, R. "Hyperhomocysteinemia and the response to vitamin supplementation." *Clin Invest* 71 (1993): 993–998.

3

From the Blues to Darkness: The World of Depression

Antidepressants, like depression, have been around for thousands of years. True, our ancient ancestors didn't have Prozac, but then again, they had a different idea of what depression is. In ancient Greece, in the time of Hippocrates, the condition we call depression was described as a mental condition involving prolonged feelings of fear and anxiety, attributed to excess "black bile." The Greek term for black bile—*melaina chole*—is the root of the English word "melancholia." Somehow calling this condition "melancholia" instead of "depression" makes it sound less, well, depressing. But no matter what you call it, depression is a major health problem in the United States. The American Psychiatric Association recognized that fact and, in 1952, developed a standard set of criteria to help identify and diagnose the disorder (see "What Is Depression?" below).

Before you learn about how SAM-e can help you or someone you care about regain his or her life from the blackness of depression, it helps to have a good understanding of what depression is, how it can affect people, and how we have been treating it, using both conventional and complementary treatments.

What Is Depression?

Carla G: "I never wanted to get up in the morning. What was the use? I didn't care about anything. It was nearly impossible to hold down a job. When I took antidepressants, I was able to work and things got better, but every drug I took made me ill, and that made me even more depressed. Then I'd stop taking the pills and the cycle would start all over again. SAM-e was my last hope."

Nicholas M: "I didn't call what I was feeling for all those years 'depression,' but I do know that I felt unhappy a lot; you know, blue. I'm sure most people thought there was nothing wrong with me at all. True, I wasn't unhappy all the time, but enough so that life just wasn't everything I hoped it would be. All that changed when I started to take SAM-e."

As these patients' remarks show, depression is a highly individual disorder that affects each person differently. Depression is a type of mood (affective) disorder. Normally, moods (e.g., sadness, joy, optimism, grief, anger) come and go, fluctuating throughout the day. When one mood dominates a person's life, it is considered abnormal. Depression is the most common mood disorder. To get an idea of how serious a problem depression is, look at the following statistics:

- More than 17 million Americans are affected by major depression each year. This is just the

number that have been formally diagnosed;
many more are undiagnosed.

- More than 28 million Americans take antidepressant or antianxiety drugs.
- Twice as many women as men are diagnosed with depression. Many experts believe the number of depressed men is higher but that male depression is underreported because men are less likely to admit they need help.
- After you've experienced one episode of depression, your chance of having another one within five years is 50 percent. After three episodes of depression, there is a 90 percent chance depression will return.
- The World Health Organization (WHO) ranks depression as the most disabling disease among women and the fourth most disabling disease overall.
- Depression costs more than $40 billion a year in lost work and health care.
- One in five people can expect to experience depression during his or her lifetime.

Some people are completely debilitated by their depressive mood; others carry on quite normally with their daily activities, or at least they appear to do so to the outside world. Some have good and bad periods, going days, weeks, or even months without feeling very low. Still others can directly link their depression to physiological changes, especially those related to hormonal fluctuations in women around menopause or after giving birth.

Everyone gets the blues or feels down or sad occasionally, even for days or weeks at a time, but that does not mean they are depressed. Many people have a tendency to use the word "depressed" to describe any kind of sad feeling. The American Psychiatric Association (APA), however, has devised eight primary criteria that more accurately describe and define depression. The symptoms are:

- Poor appetite accompanied by weight loss, or an increase in appetite accompanied by weight gain
- Lack of energy and feeling fatigued
- A decreased ability to think clearly or to concentrate
- Feelings of worthlessness, hopelessness, or inappropriate guilt
- Inability to sleep or wanting to sleep all the time (insomnia or hypersomnia)
- Loss of interest or enjoyment in activities that once were pleasing, and/or loss of sexual desire
- Physical inactivity or hyperactivity
- Persistent or recurrent thoughts of death or suicide

According to the APA, anyone who has five of these eight symptoms for at least one month has clinical depression (also known as major depression or unipolar depression). Individuals with any four symptoms for the same length of time are described as being simply depressed.

Mild Depression

Not all depression is the same, so the APA established guidelines to help recognize each of the forms of depression. One of these forms is dysthymia, or mild depression. A person is diagnosed with dysthymia if he or she has been depressed for most of the time for at least two years (one year for children or adolescents) and has at least three of the following symptoms:

- No self-confidence or low self-esteem
- Feelings of despair, hopelessness, or pessimism
- Fatigue or lethargy
- Lack of interest or pleasure in ordinary activities
- Reduced productivity
- Difficulty making decisions or concentrating
- Isolation from social activities
- Guilt or excessive thoughts about the past
- Excessive anger or irritability

Bipolar Depression

In contrast to mild depression there is bipolar (manic) depression. Only 5 percent of people with depression have bipolar depression; all other types are called unipolar depression. A person with bipolar depression experiences alternating episodes of major depression and elevated mood. The APA lists the following symptoms to identify bipolar depression:

- Excessive self-esteem or feelings of grandiosity
- Feeling that the mind is racing; thoughts run rapidly through the mind
- Inability to concentrate; easily distracted
- Reduced need for sleep
- Extremely talkative and an obsessive need to talk
- Inability to exercise good judgment, resulting in excessive actions such as uncontrolled spending, unwise financial decisions, and inappropriate sexual behaviors
- Increase in work or social activities, often involving sixty- to eighty-hour work weeks

Varieties of Depression

Depression can be situational, affecting people at specific stages of their lives or at different times of the year, and then disappearing. One such variety of depression affects women who have just given birth. Called **postpartum depression**, it affects up to 80 percent of new mothers and usually disappears within two weeks of delivery. This is about the time it takes for a woman's hormones to return to their pre-pregnancy levels. For up to 15 percent of these new mothers, however, the crying spells, mood swings, and fatigue that characterize this depression continue on and on, and some of these women go on to develop bipolar depression. Women who are considering taking SAM-e for postpartum depression should be diagnosed by a physician to rule out bipolar depression before taking the supplement.

Another variety of depression that affects women only and which has been linked to hormonal changes is **postmenopausal depression**. The decline in estrogen levels along with lifestyle changes are believed to contribute to this depression.

For some people, where they live can trigger depression. People who suffer with SAD, or **seasonal affective disorder,** get depressed during the autumn and winter months only, when the daylight hours are short and the hours of darkness are long. Symptoms include lethargy, a tendency to cry easily, overeating, and wanting to sleep a lot. The symptoms go away when spring comes.

Depression: A Whole Body Experience

As all the criteria for the different types and varieties of depression indicate, depression is a whole body experience that affects the mind, body, and spirit. Each of these symptoms can involve many different behaviors and emotions that can impact every part of a person's life. A loss of interest in activities, for example, may mean a person withdraws from family and friends or is unable to cooperate with co-workers. An inability to think clearly may make a person who was once even-tempered become angry and irritable. Some depressed people cry a lot; others have panic attacks, provoke arguments, or become afraid to go out. Some try to find comfort in drugs, alcohol, food, or tobacco, which only make their situation worse.

Before you get caught up in counting how many symptoms you have and trying to decide if you're depressed or not, remember that depression is a personal, subjective experience. The death of a spouse may cause one surviving partner to sink into a depression he or she never comes out of, while another may grieve for many months yet come to peace with the situation and go on to live and love again. Criteria are guidelines only; only you know how you feel.

SAM-e has been found to be an effective treatment for all types of depression except bipolar depression. That means 95 percent of people who are experiencing some degree of depression, whether it's occasional blues, moderate, recurrent episodes, or chronic clinical depression, may get fast, effective help with SAM-e. (People with bipolar depression should not use SAM-e or any other antidepressant unless they are under a doctor's care because it can exacerbate manic episodes. If you are experiencing chronic depression, seek professional help before self-treating with SAM-e.) A detailed discussion of the use of

SAM-e in the treatment of depression is covered in Chapter 4.

Who Is at Risk for Depression?

Depression can affect a one-month-old infant or a one-hundred-year-old man, and all ages in between. Stressors come in all forms and at all stages of life. For an infant, the stress can come in the form of a neglectful mother. People who have chronic health problems are often depressed and usually do not recognize it. Many new mothers experience temporary postpartum blues, but for about 15 percent of all new mothers, the blues linger on and on. Men and women who are overwhelmed by the demands of family, school, and career can become depressed. And among the elderly, depression is often written off as normal aging.

Three groups of people have an especially hard time dealing with depression: children, adolescents, and the elderly. Depression in these vulnerable individuals deserves a closer look.

Depression in Children

Depression usually first appears during childhood or adolescence, yet it also can make an initial appearance in infancy. Depression among infants manifests as listlessness, refusal to eat, and lack of normal development (known as *failure to thrive*). The younger a child is when he or she has the first episode, the more likely the child is to have recurring depression throughout his or her lifetime. As many as 10 percent of children younger than age thirteen may be suffering with depression. There have been cases of children as young as five years old committing suicide.

Identifying depression in children is a bit different from identifying it in adults. Here are some symptoms to look for. If they continue for two weeks or longer, take the child to a physician to rule out any medical problem that could be causing the depression, and seek help from a child psychiatrist if no medical condition is found. Watch to see if the child:

- Is pessimistic and negative
- Is lethargic or reports having no energy
- Remains sad even when involved in activities that used to be a source of joy
- Is having trouble in school
- Avoids friends
- Talks about death or harming him- or herself
- Is having sleep problems or nightmares
- Looks unhappy most of the time
- Often reports feeling sad or bored or sick

Depression in Adolescents

Adolescence is a difficult stage of life, and rapidly changing moods are not uncommon. However, in an estimated 5 to 10 percent of teens mood changes and other symptoms are an indication of depression or manic depression. Less than half of these adolescents get professional help, and up to 55 percent of depressed teenagers elect alcohol or drugs as a way to deal with their feelings. And for those teens who can't take the depression any longer, suicide is often chosen as a way out. Suicide is the third leading cause of death among this age group, behind homicides and accidents.

To help spot signs of depression in a teenager, look to see if the child:

- Stops participating in activities he or she once enjoyed
- Grows distant from family and friends
- Changes in appearance—dirty clothes and hair
- Is aggressive, angry, or easily irritated most of the time, without cause
- Begins to smoke or use drugs or alcohol
- Becomes sexually promiscuous
- Sleeps a lot or very little
- Takes part in dangerous and/or criminal activities
- Has difficulty remembering things
- Talks about death or hurting him- or herself or is preoccupied with death

If your teenager is depressed, get help immediately. Have your physician determine if there is a medical cause, and get a referral for a psychiatrist who specializes in adolescents. Discuss with the health professionals whether SAM-e may be able to help your child. (See Chapter 9 for how to use SAM-e.)

Older People and Depression

Depression among the elderly is a widespread and underrecognized problem. According to a report in *Clinical Geriatric Medicine*, 15 percent of the elderly who live in the community have depressive symptoms, while another source adds that the figure among outpatient medical clinics is 30 to 50 percent and in nursing homes and other institutions it is 25 percent.

Perhaps just as distressing as these high percentages is the fact that up to 90 percent of depressed people sixty-five years and older fail to get any kind of help with their depression. This lack of attention is due largely to two factors that are somewhat related: a misconception among many people that depressed mood is a normal response to aging; and the fact that many family members and physicians simply miss the symptoms, especially when individuals have other, physical ailments.

A recent study looked at the idea that growing old and depression go hand-in-hand. The investigators evaluated 2,219 older people over a two-year period to determine if aging is an independent risk factor for depression. They concluded that age-related situations, such as chronic health problems and other disabilities, increase the risk of developing depression. Among healthy, normally functioning older adults, however, the risk for becoming depressed was no greater than for younger adults.

Aging is accompanied by numerous physical, psychological, and social changes, many of which can cause or contribute to depression. These changes can be summed up in one very powerful word—loss. Losses can include job, health, finances, loved ones (spouse, family members, friends), home (because of relocation, forced or otherwise), and social status. Many elderly people feel they have lost control of their lives, especially when they have lost their health and must depend on others for their daily care. Because these changes occur, the diagnosis and treatment of depression in older people can be a challenge for physicians.

As mentioned earlier, another problem with detecting depression in the elderly is that it is often attributed to other medical conditions. Signs of depression are common in people who have hypothyroidism, diabetes, early Alzheimer's disease, stroke, and Parkinson's disease, as well as individuals who are taking medications, like antihypertensives and heart medications, which can cause depressive mood swings (see page 44 for drugs that can cause depression). In such cases, the depression is often ignored and left untreated.

Signs of depression in the elderly often are apparent in their behavior. Older people who are depressed may stop taking care of their appearance or stop taking their prescribed medications. Some will complain persistently about pain or fatigue while others may keep going to doctors in search of medical care despite the lack of physical symptoms. Very often clues to depression are apparent in their mood or words. They may be uncharacteristically irritable, have feelings of hopelessness and helplessness, talk about wanting to die or have suicidal thoughts, and become very impatient with people and situations.

Proper treatment of depression among the elderly is critical. (See Chapter 4 for information on treatment of depression in the elderly.) According to research published in the *Journal of the American Geriatric Society* in 1996, depression among the elderly, left untreated, can lead to increased mortality from suicide as well as increased complications and mortality from medical conditions. The National Center for Health Statistics (1992) reported that while suicide rates for all ages is about 12.2 per 100,000, the rate among the elderly is nearly twice that rate: 20.1 per 100,000. Among white elderly men it is especially high—40.7 per 100,000.

An example of the impact of untreated depression among the elderly is evident in a study reported in *Mental Health Weekly*. Following more than ten thousand elderly men and women with hypertension (high blood pressure) over a three-year period, researchers found that those with depression had strokes at up to 2.7 times the rate of those without depression. According to Eleanor Simonsick, M.D., an epidemiologist at the National Institute on Aging, "If elderly people have depression, it complicates their medical risk across the board." One reason for increased risk is that depressed individuals are less likely to take care of themselves; for example, eating a healthy diet, getting enough exercise, socializing, and taking needed medications.

Problems with Diagnosing Depression

Even though the APA has criteria for various types of depression, even though people talk about depression, it is not an easy illness for doctors to diagnose. First, they lack accurate tools: there are no quick, definitive tests they can use to identify depression—no simple blood test, no fancy X rays. Unless they are specially trained in psychiatry or have experience with treating depressed patients, they are not likely to recognize it. Second, many people who are depressed go to their doctor complaining of symptoms such as fatigue, weight gain, insomnia, or pain, all of which can be caused by depression, yet the doctor usually gives a prescription to treat the symptoms rather than search for the cause. Third, unless depressed individuals specifically tell their doctor they are experiencing depression, it is virtually impossible for the physician to detect it during a routine office visit. Unfortunately, depressed individuals, especially those who are most depressed, are the least likely to tell their doctor they need help with depressed mood.

Sometimes depression is the symptom rather than the primary condition (see "Causes of Depression" below). Many medical problems have depression as one of their main symptoms, including arthritis, cancer, chronic pain, diabetes, heart disease, liver disease, lung disease, multiple sclerosis, and Parkinson's disease. In these cases two different problems can arise: the primary condition is treated but the depression is not; or the depression is treated and the primary condition is missed. Therefore it is important for everyone who is experiencing depression, whether it is mild or more serious, to have a thorough physical examination to determine if depression or another medical condition is the primary problem. Failure to address an underlying cause of depression will make any treatment of the depression less effective. In addition, delaying treatment of an organic cause of depression may also allow that condition to worsen unnecessarily.

Depression Is Misunderstood

Many people in American society, including physicians, have the misconception that people who are depressed can simply "get over it" if they want to. It's as if they believe depression is a choice; that depressed people want to feel miserable, lack joy in their lives, and feel they are unloved and unwanted. Part of the problem may be the very word itself, says neuroscientist Philip Gold, who is with the National Institutes for Mental Health. "People confuse it with the everyday sensation of feeling despondent and dismiss it. In fact, it takes an incredibly strong person to bear the burden of the disease."

Depression is an illness, a physical disorder that causes chemical changes in the brain. Wishing that these chemicals will correct themselves is just that: wishful thinking. In order to treat depression effectively, you need to correct the chemical imbalances in the brain and, in many cases, learn how to better cope with those stressors in life that could trigger those chemical imbalances again.

If left untreated, the effects of depression can be very damaging to a person's health, relationships, job, and future. This is true for mild as well as for major depression, because untreated mild depression can progress into major depression. Over time, the effects of depression accumulate and may cause some people to lash out at others, physically, verbally, or both, or to turn their anger inward. It is estimated that 15 percent of all people who have major depression will commit suicide. About 80 percent of people who commit suicide talk about it before they commit the act, although often their words are disregarded by people around them. Among children and adolescents who commit suicide, drawings and writings about suicide are a common clue.

Causes of Depression

Since World War II, the incidence of depression has nearly doubled. The cause of this increase, says Myrna Weissman, an epidemiologist at Columbia University, is not only a greater public awareness of the condition, but also an increase in stress, a disintegration of family and social support, and a rise in nutritional deficiencies. Actually the list of causes of depression is quite long, yet they all have one thing in common: they all ultimately involve an imbalance in the biochemistry of the brain, whether that imbalance is triggered by drugs, a nutritional deficiency, stress, hormone changes, allergies, a hereditary vulnerability that results in depression when stimulated in a certain way, or any one of the other listed causes. The brain, however, is where the action takes place.

Depression and the Brain

Depression begins and ends in the brain. Here's a simplified version of how depression works: The brain contains billions of nerve cells called neurons. Neurons consist of a central body from which project many spindle-like branches called dendrites. Neurons are constantly communicating with one another, and to do so they pass messages or signals along the dendrites, which are separated from one another by tiny spaces called synapses. So when a signal needs to go from one neuron to another, it releases a chemical called a neurotransmitter, which elevates mood and which must leap across the synapse from one dendrite and attach itself to a receiving dendrite of another neuron. This action triggers the release of more neurotransmitters on the receiving neuron. Immediately after these neurotransmitters are released, enzymes go to work on the original neurotransmitter and either break it down or help it back into the neuron that produced it so it can be used again. When the level of any of these neurotransmitters is deficient, mood lowers and depression can result.

What does this process have to do with SAM-e and depression? The keys to the process are the neurotransmitters—serotonin, dopamine, and noradrenaline (or norepinephrine). SAM-e is responsible for transferring methyl molecules to these neurotransmitters in the brain and boosting their levels. An imbalance of any one or more of these chemicals (which collectively are in a class known as monoamines) can cause depression.

Once researchers realized the connection between neurotransmitter activity and depression, they began to develop drugs that helped maintain or restore neurotransmitter levels. A more detailed explanation of how these pharmaceutical drugs affect neurotransmitter activity and the side effects they cause is discussed in Chapter 4.

Stress and Depression

Few people realize the tremendous impact stress has on the body and its brain chemistry. And because every person reacts differently to stressful situations, it is sometimes hard for people to understand why one person becomes depressed over a situation while another person breezes through the same situation without any apparent problem. Whether you eventually become depressed from exposure to stress and to what degree you will be depressed depend on many factors, including how well you have developed coping mechanisms and what particular inherited susceptibilities you have for handling stress.

Stress can trigger numerous reactions in the body that can cause depression. For example, significant emotional stress can alter the function of the adrenal gland, causing it to release excessive amounts of a hormone called cortisol into the system. At high levels, cortisol can be an especially destructive hormone. Besides increasing the risk of stroke, osteoporosis, diabetes, and heart disease, it also interferes with the activity of the neurotransmitters and in extreme cases can cause schizophrenia.

Nutrition and Depression

The delicate biochemical balance in the brain can be easily altered, and the result can be depression when even one essential nutrient is deficient. SAM-e has an intimate relationship with two vitamins proven to cause depression when they are at low levels in the body—folic acid and vitamin B_{12}. (Vitamin B_6 is also important, but less so, and is discussed elsewhere.) In studies of depressed patients, between 31 and 35 percent were found to have a deficiency of folic acid. These figures were dramatically higher in several other studies. In one, up to 92.6 percent of elderly patients admitted to a psychiatric facility had a folic acid deficiency. When depressed individuals with either of these deficiencies are given supplements, dramatic improvements have been achieved.

Until recently, folic acid deficiency was believed to be the most common nutrient deficiency in the world. Although the situation is improving slowly as people are becoming aware of this problem and more and more prepared foods are being enriched with folic acid, it is still a very significant problem. Vitamin B_{12} deficiency is less common than folic acid deficiency, but it still remains a cause of depression, especially among the elderly.

SAM-e needs a little help from folic acid and vitamin B_{12} to relinquish its methyl molecules. All three substances—SAM-e, folic acid, and vitamin B_{12}— transfer methyl molecules to compounds in the brain, including neurotransmitters. All three need to be in sufficient supply to perform methylation. This means you may need to supplement these nutrients while taking SAM-e (see Chapter 9). Without the contribution of the methyl molecule from these three donors, the neurotransmitters cannot function properly, their levels decline, and depression can set in.

Another compound in the brain that depends on methylation is tetrahydrobiopterin (BH_4). This compound is involved in the manufacture of the neurotransmitters that have a role in depression, which include norepinephrine, serotonin, and dopamine. Studies show that patients who experience recurrent depression have reduced synthesis of BH_4, which scientists believe is related to low levels of SAM-e.

Thus there is a close relationship among all of these compounds—SAM-e, folic acid, vitamin B_{12}, and BH_4—in causing depression. Several other nutrients also have an important role in depression, especially members of the B complex vitamins—thiamin, riboflavin, niacin, biotin, pantothenic acid, and vitamin B_6—as well as vitamin C. Information on how to supplement with these vitamins when taking SAM-e is discussed in Chapter 9.

Drugs: Prescription, Nonprescription, Other

Depression is a common side effect of many prescription and nonprescription drugs, as well as alcohol, tobacco, and caffeine. All of these drugs deplete the body of essential nutrients or substances that affect mood.

Although many people consider alcohol to be a stimulant, it is actually a depressant. It increases the secretion of the hormone cortisol from the adrenal gland (see "Stress and Depression" above), disrupts brain cell functioning, and disturbs sleep patterns. Similar to alcohol, nicotine stimulates the adrenal gland, increasing its output of cortisol. Caffeine is a stimulant, although not everyone responds to the same degree. Individuals who tend to feel depressed usually react the most, experiencing depression, heart palpitations, nervousness, and headache with caffeine intake. In several studies among healthy college students and depressed patients, researchers found that high intake of caffeine is associated with greater or more severe depression.

Food Allergies

An association between food allergies and depression was first recognized by the scientific community in

---

Drugs That May Cause Depression

Ace inhibitors (e.g., enalapril, fosinopril)

Ammonium chloride

Amphetamines (e.g., dextroamphetamine)

Analgesics (e.g., diflunisal, etodolac, fenoprofen, flurbiprofen, ibuprofen, propoxyphene)

Antiarrhythmics (e.g., digoxin, procainamide)

Antihistamines

Antihypertensives (e.g., atenolol, captopril, diltiazem, furosemide, nadolol, terazosin)

Anti-inflammatory agents (e.g., diclofenac)

Baclofen

Barbiturates

Beta-blockers (e.g., carteolol, timolol)

Diuretics (e.g., triamterene)

Estrogens and oral contraceptives

Tranquilizers and sedatives

1930, when allergist Albert Rowe, M.D., described an allergic condition that included depression, fatigue, drowsiness, and muscle and joint pain. No one uses Dr. Rowe's term, "allergic toxemia," anymore, but food allergies are still recognized as a cause of depression. Other symptoms that often accompany food allergies include chronic fluid retention, puffiness under the eyes, sinus congestion, and chronic swollen glands. Foods most often associated with allergic reactions include milk, wheat, eggs, chocolate, and oranges, as well as food additives, preservatives, and artificial colorings and flavorings found in many processed food products.

Environmental Toxins

Many environmental poisons, including pesticides and herbicides, heavy metals (e.g., mercury, arsenic, aluminum), and solvents used in manufacturing and cleaning (e.g., formaldehyde, benzene) are attracted to nerve cells. Exposure to these substances, especially chronic exposure, can set off an explosion of free-radical activity and cause various dysfunctions of the central nervous system, including depression, confusion, and headache. The potent antioxidant glutathione, which cannot be produced without SAM-e, protects against free-radical damage (oxidation).

Preexisting Medical Conditions

People who have various medical conditions, especially chronic ones (e.g., diabetes, multiple sclerosis, Parkinson's disease) are often depressed. In Parkinson's disease, for example, one study indicated the incidence of depression is about 46 percent while another showed that 32 percent of patients with Parkinson's had been depressed for most of their lives.

When looking at depression and disease, it is not always clear whether the depression is a symptom or a consequence of the medical condition (see "Problems with Diagnosing Depression" above). As the APA criteria for depression show, physical complaints are part of the depression picture. Physical disorders that are often associated with depression are listed in the sidebar below. Regardless of whether the depression is a symptom or a consequence, SAM-e has proved useful in improving the quality of life in patients with chronic disease. Some of the studies that have looked at this problem are discussed in Chapter 4.

Causes of Depression: The Dilemma

There is much about depression that scientists do not understand, and that includes the causes of depression. Often it is a matter of which came first, the chicken or the egg. Does a decline in neurotransmitter levels cause depression, or does depression cause the levels to drop? Does emotional stress trigger a depressive mood, or does the depressive mood trigger emotional stress? Depression is not always a case of

---

Physical Diseases and Ailments That Are Associated with Depression

Alzheimer's disease and other forms of dementia; Cancer; Chronic pain; Diabetes; Epilepsy; Heart disease; Hepatitis; Huntington's disease; Liver disease; Lung disease; Lyme disease; Multiple sclerosis; Parkinson's disease; Rheumatoid arthritis; Viral infections (e.g., flu, mononucleosis, pneumonia, viral meningitis); Wilson's disease

black and white; there are many gray areas still to be explored.

It is worth repeating that while antidepressants, including SAM-e, can relieve the symptoms of depression, they cannot address its source. That's why most experts recommend that people who are moderately to severely depressed seek counseling to help them deal with the cause of their depression. It is not within the scope of this book to discuss the various counseling and psychological therapy options for dealing with depression. A list of resources for you to contact regarding such services is given in the appendix.

Mildly depressed individuals who turn to antidepressants usually find that the symptomatic relief they get is all they need and that professional psychotherapy may not be necessary. In any case, the antidepressant you choose, after consulting with your health-care provider, to help you resolve your depressive mood should be one that is safe, effective, and without debilitating or distracting side effects. SAM-e offers that promise to you.

NOTES

Abalan F., et al. "Frequency of deficiencies of vitamin B_{12} and folic acid in patients admitted to a geriatric-psychiatry unit." *Encephale* 10 (1984): 9–12.

The American Psychiatric Association. *Diagnostic and Statistical Manual of Mental Disorders (DSM-IV)*.

Bolton, S., and Null, G. "Caffeine, psychological effects, use and abuse." *Journal of Orthomolecular Psychiatry* 10 (1981): 202–11.

Breggin, P. R., and Breggin, G. R. *Talking Back to Prozac: What Doctors*

Aren't Telling You About Today's Most Controversial Drug. New York: St. Martin's Press, 1994.

Brostoff, J., and Challacombe, S. J., ed. *Food Allergy and Intolerance*. Philadelphia, PA: W. B. Saunders, 1987.

Carney, M. W. P., et al. "Red cell folate concentrations in psychiatric patients." *Journal of Affective Disorders* 19 (1990): 207–213.

Chou, T. "Wake up and smell the coffee: Caffeine, coffee, and the medical consequences." *West J Med* 17 (1992): 544–553.

Conwell. "Outcomes of depression." *Journal of the American Geriatrics Society* 4(Suppl. I) (1996): S34–44.

Crellin R., Bottiglieri, T., and Reynolds, E. H. "Folates and psychiatric disorders: Clinical potential." *Drugs* 45 (1993): 623–636.

Curtius, H., Muldner, H., and Niederwieser, A. "Tetrahydrobiopterin: Efficacy in endogenous depression and Parkinson's disease." *J Neural Trans* 55 (1982): 301–308.

Curtius, H., et al. "Successful treatment of depression with tetrahydrobiopterin." *Lancet* (1983): 657–659.

Gilliand, K., and Bullick, W. "Caffeine: A potential drug of abuse." *Adv Alcohol Subst Abuse* 3 (1984): 53–73.

Godfrey, P. S. A., et al., "Enhancement of recovery from psychiatric illness by methylfolate." *Lancet* 336 (1990): 392–395.

Hamilton, M. S., Opler, L. A. "Akathisia, suicidality, and fluoxetine." *Journal of Clinical Psychiatry* 53(11) (November 1992): 401–406.

Hughes, Ellen. "Depression: Living with and treating the disorder." The American Medical Association Web site.

Kalda, R. "Media- or fluoxetine-induced akathisia." *American Journal of Psychiatry* 150(3) (March 1994): 531–532.

Katz, I. R., Curlik, S., and Nemetz, P. "Functional psychiatric disorders in the elderly." In L. W. Lazarus, ed. *Essentials of Geriatric Psychiatry*. New York: Springer, 1988.

Kivela, S. I., Pahkala, K., and Eronen, A. "Depression in the Aged: Relation to folate and vitamins C and B_{12}." *Biol Psychiatry* 26 (1989): 209–213.

Koenig, H. G., and Blazer, D. G. "Epidemiology of geriatric affective disorders." *Clinical Geriatric Medicine* 8 (1992): 235–241.

Lipinski, J. F., Mallya, G., Zimmerman, P., Pope, H. G. "Fluoxetine-induced akathisi. Clinical and theoretical implications." *Journal of Clinical Psychiatry* 59(9) (September 1989): 339–342.

Mental Health Weekly 5(37) (25 September 1995): 7.

Murray, Michael T. *Encyclopedia of Natural Medicine*. Rocklin, CA: Prima Publishing, 1998.

Murray, Michael T. *Natural Alternatives to Prozac*. New York: William Morrow, 1996.

National Center for Health Statistics, 1992. "Advance report of final mortality statistics, 1989." NCHS Monthly Vital Statistics Report 40 (8 Suppl 2).

Neil, J. F., et al. "Caffeinism complicating hypersomnic depressive disorders." *Comprehensive Psychiatry* 19 (1978): 377–385.

Peterson, C. "Explanatory style as a risk factor for illness." *Cognitive Therapy and Research* 12 (1988): 117–30.

Power, A. C., and Cowen, P. J. "Fluoxetine and suicidal behavior. Some clinical and theoretical aspects of a controversy." *British Journal of Psychiatry* 161(12) (December 1992): 735–741.

Reynolds E., et al. "Folate deficiency in depressive illness." *British Journal of Psychiatry*117 (1970): 287–292.

Roberts, R. E., et al. "Does growing old increase the risk for depression?" *American Journal of Psychiatry* 154 (October 1997): 1384–1390.

Rowe, A. H., and Rowe, A., Jr. *Food Allergy: Its Manifestations and Control and the Elimination Diets: A Compendium.* Springfield, IL: Charles C. Thomas, 1972.

Sabaawi, M., Holmes, T. F., and Fragala, M. R. "Drug-induced akathisia: Subjective experience and objective findings." *Military Medicine* 159(4) (April 1994): 286–291.

Schufte, J. L. "Homocide and suicide associated with akathisia and haloperidol." *American Journal of Forensic Psychiatry* 6(2) (1985): 3.

Thornton, W. E., and Thornton, B. P. "Geriatric mental function and folic acid: A review and survey. *Southern Medical Journal* 70 (1977): 919–922.

US News & World Report, "Melancholy nation," 126(9) (March 8, 1999): 56.

Wirshing, W. C., Van Putten, T., and Rosenberg, J. "Fluoxetine, akathisia and suicidality. Is there a causal connection?" *Archives of General Psychiatry* 49(7) (July 1992): 580–581.

Zucker, D., et al. "B_{12} deficiency and psychiatric disorders. A case report and literature review." *Biol Psychiatry* 16 (1981): 197–205.

4

Conventional and Alternative Treatments for Depression: The Promise of SAM-e

If you or someone you care about is living with depression, you know how important it is to find a good antidepressant. And if you are like most people, you also know how hard it is to find one that works, that lasts, and that doesn't cause side effects that are so terrible they cause you to quit treatment.

This chapter takes a close look at the antidepressants from which you can choose. Naturally, the headliner of the chapter is SAM-e, for reasons already hinted at and others that will become clear once you read more about how this supplement works for depression. But before you learn more about SAM-e, it is important for you to understand how the antidepressant drugs on the market work. You may be taking one of these drugs right now, or perhaps you've taken one or more of them in the past. You may even be thinking about taking one but want to learn about your options before you do.

In this chapter you'll be introduced to SAM-e and the studies that support its use for depression. You will see how it stacks up against antidepressive drugs and how it is superior to them. The chapter closes with a look at some other natural antidepressants that, although not as powerful as SAM-e, are effective and can serve as complementary treatments with SAM-e.

Pharmaceutical Treatment for Depression

Scientists know that to treat depression effectively, the chemical balance in the brain must be restored. To accomplish that, many people need to take some type of antidepressant (pharmacological or natural) that affects the brain chemicals involved in depression. A large number of people, however, get no relief from conventional antidepressants. About one-third of people do not respond to antidepressants at all, while for others the drugs work for a while and then stop providing relief. Between 20 and 30 percent of people who enter antidepressant studies drop out because the side effects are too disturbing or disruptive to their lives. Obviously, antidepressants are not for everyone, which is one reason why some people have been turning to natural alternatives, such as St.-John's-wort and 5-HTP, to treat mild depression.

But some people need to continue taking antidepressants for the rest of their lives unless they learn ways to deal with their depression by participating in some sort of counseling, therapy, or self-help program that allows them to develop coping skills. Most experts recommend that people with any significant degree of depression undergo some type of counseling anyway, which for some will completely eliminate the need for drugs and for others will reduce the dose while enhancing their quality of life.

Types of Pharmaceutical Antidepressants

Four types of pharmaceutical antidepressant drugs are currently used to treat depression. Each of these categories is discussed below, with an explanation of how the drugs work (to the best of our knowledge, because no one has yet identified exactly how these drugs work) and the side effects associated with their use. Another class of drugs, the benzodiazepines, are also used to treat depression and are discussed as well. Regardless of the type of pharmaceutical you take to treat depression, there are some general caveats you need to consider.

1. Never combine different types of antidepressants, an antidepressant and tryptophan (an amino acid that is a precursor to serotonin), or antidepressants and SAM-e without first consulting your physician. In the first case, you can experience serious drug reactions; in the second, tryptophan (an amino acid) can enhance the side effects associated with antidepressants (for example, if combined with Prozac, it can increase restlessness, gastrointestinal symptoms, and agitation); in the third, combining SAM-e with a pharmacological antidepressant can alter the onset of the response to the drug.

2. Do not suddenly stop taking an antidepressant if you have been taking it for two months or more without first checking with your doctor. Abrupt withdrawal from conventional antidepressants can trigger severe reactions, including

palpitations, tremors, low blood pressure, and other serious symptoms.

3. The estimated effectiveness of tricyclic antidepressants and selective serotonin reuptake inhibitors (SSRIs) is 70 percent. This figure is inflated, however, because it does not take into account the high number of people who drop out of the drug studies because of intolerable side effects. Thus the 70 percent figure only reflects the people who actually finish the studies. What the real figure is, is anyone's guess.

Tricyclic Antidepressants

The first tricyclic antidepressant was introduced to the marketplace in the 1960s. Tricyclic antidepressants get their name from their chemical structure, which consists of three (tri) rings (cyclic). Researchers believe they relieve depression by preventing the nerves from trying to take back (or reuptake) two of the neurotransmitters associated with depression, serotonin and norepinephrine. Thus tricyclic antidepressants allow these neurotransmitters to remain in service.

The most commonly prescribed tricyclics are amitriptyline (Elavil, Endep), desipramine (Norpramin, Pertofrane), doxepine (Adapin, Sinequan), imipramine (Imavate, Presamine, Tofrinil), and nortriptyline (Aventyl, Pamelor). All are available in generic as well as trade (brand) names (shown here in parentheses). It takes at least three weeks, and often longer, for tricyclics to begin to express their antidepressant effect. Unfortunately, the undesirable side effects can begin earlier. The most common are dry mouth, constipation, weight gain or loss, nausea, a "drugged" feeling, blurry vision, lightheadedness, fainting, and urinary retention. Both men and women can experience loss of sexual desire and a change in the size of their breasts, while men may also have testicular swelling. Less common side effects include rash, hives, and itching. Among older people, urinary retention and loss of bladder control (urinary incontinence) can occur among both sexes. Because tricyclics can cause lightheadedness and fainting, their use should be carefully monitored in the elderly to help avoid falls and injury.

A common problem that occurs with tricyclics is overdosing. According to the Substance Abuse and Mental Health Services Administration, 53 percent of drug-related admissions to emergency rooms in the United States are due to drug overdose. Many people overdose on tricyclics because they take too many pills between the time they start drug therapy and when the drug begins to take effect. In 1994, 90 percent of emergency room visits related to tricyclic antidepressants occurred because patients had overdosed, either intentionally or unintentionally.

On rare occasions, tricyclics can cause more serious side effects, including hepatitis and jaundice. A potentially fatal reaction associated with tricyclic use is irregular heart rhythms. Many physicians avoid giving tricyclics to individuals who have a heart condition, although studies show doxepine is safer than the other drugs in the group.

Some physicians support the use of SAM-e along with tricyclics to enhance the effects of the drug. This combination, which should be done only under the supervision of your physician, can reduce the number and severity of side effects and reduce the tricyclic dose.

Selective Serotonin Reuptake Inhibitors (SSRIs)

This group of antidepressants is probably best known for its "star," fluoxetine, or Prozac. The SSRIs work similarly to the tricyclics, except that the SSRIs inhibit the reuptake of serotonin alone (hence their name). They also have a wider range of conditions for which they are reported to be effective, including obsessive/compulsive disorders, schizophrenia, and eating disorders.

The SSRIs broke into the U.S. market in 1988, more than thirty years after researchers discovered the neurotransmitter serotonin in the brain. Originally the reports were that SSRIs caused much fewer side effects than tricyclics. However, since then it's been found that they are responsible for more side effects than originally thought, although many experts still consider SSRIs to be an improvement over tricyclics. Prozac is the most controversial of the SSRIs, and is discussed briefly below.

Since the introduction of Prozac, other SSRIs have been developed and joined the group, including Celexa (citalopram), Luvox (fluvoxamine), Paxil (paroxetin), and Zoloft (sertraline). The only SSRI member that has a slightly different action is Zoloft, which also increases the level of dopamine in the brain.

At least 20 percent of patients who take SSRIs experience nausea, headache, and insomnia. Around 10 percent can expect anxiety, drowsiness, diarrhea, weakness, loss of appetite, dry mouth, nervousness, tremors, stomach pain, and sweating. Less common reactions include sore throat, lack of sex drive, muscle pain, rash, flatulence, fever, and palpitations.

The controversy surrounding the use of Prozac deserves brief mention here. Readers who want more information can refer to the Notes at the end of this chapter.

According to Peter R. Breggin, M.D., the FDA psychiatrist who was involved in the agency's safety review of Prozac and the author of *Talking Back to Prozac: What Doctors Aren't Telling You About Today's Most Controversial Drug*, Prozac stimulates the nervous system and thus can produce highly undesirable effects similar to those caused by amphetamines ("speed"). Two of those effects are akathisia (an uncontrollable need to pace and otherwise move around) and extreme agitation. Although the manufacturer of Prozac claims the drug may cause akathisia in less than 1 percent of people who take it, a report in a 1989 issue of *Journal of Clinical Psychiatry* found the figure was between 10 and 25 percent. These high figures are supported by other studies as well, including those published in the *Archives of General Psychiatry* and *American Journal of Psychiatry*.

Researchers have also found links between use of Prozac and violent and suicidal acts and/or thoughts. In one report in the *Journal of Forensic Psychiatry*, three patients who were taking Prozac and who were experiencing akathisia either attacked other people or committed murder. In two other studies (*British Journal of Psychiatry* and *Journal of Clinical Psychiatry*) researchers described patients on Prozac who became preoccupied with suicidal thoughts. Defenders of Prozac say that some people will commit suicide or have suicidal thoughts regardless of the antidepressant they are taking.

The use of Prozac remains controversial. Millions of people around the world take the drug, most of them with apparent safety. Given its high profile, investigations into its safety will continue until more definitive answers are found.

Monoamine Oxidase Inhibitors

The drugs in this class inhibit the action of an enzyme called monoamine oxidase, which breaks down the neurotransmitters in the brain. By inhibiting this enzyme, the drugs enhance the levels of serotonin, dopamine, and norepinephrine. Monoamine oxidase inhibitors (MAO inhibitors) were first developed in the 1950s.

Use of MAO inhibitors is usually reserved for people who have severe depression or panic disorders. They are very potent drugs and are associated with disturbing side effects, including sexual dysfunction, fluid retention, insomnia, headache, anxiety, fatigue, dizziness, and weight gain. Because dosing with MAO inhibitors can be tricky, and fatal if handled improperly, they should only be taken under the close supervision of a physician. Common MAO inhibitors include phenelzine (Nardil) and tranylcypromine (Parnate).

Anyone taking MAO inhibitors also needs to be aware of the long list of drug interactions. Neither SAM-e nor any other antidepressant should be used if you are taking MAO inhibitors. For at least two weeks after discontinuing MAO inhibitors, avoid using any over-the-counter cold remedies, including any that contain the herb ephedra or ma huang, and any prescription drugs. MAO inhibitors can even react with foods. Because the tyramine in certain foods can cause extremely elevated blood pressure in people who take MAO inhibitors, the following foods should be avoided: anything that contains alcohol (including cough syrups); coffee, broad beans, and chicken livers; and aged or fermented foods such as cheese, yeast, processed meats, bean curd, soy sauce, and pickled foods.

Newer Antidepressant Drugs

Several other antidepressant drugs do not fit into any of the above categories. These are among the newer antidepressants on the market and include Desyrel (trazodone), Effexor (venlafaxine), Remeron (mirtazapine), Serzone (nefazodone), and Wellbutrin and Zyban (bupropion). Bupropion is the most widely prescribed of all these drugs, even though it has been associated with seizures when administered in high doses. All of these antidepressants typically are associated with fewer side effects than those found in the other categories, including an absence of sexual problems; nausea, diarrhea, and insomnia are the major complaints.

Benzodiazepines

Perhaps the most well-known drug in this category is Valium. Benzodiazepines are tranquilizing drugs most often used to treat anxiety and insomnia, but often prescribed by primary care physicians, alone or along with Prozac, to treat depression. Examples of benzodiazepines include alprazolam (Xanax), chloazepate (Tranxene), chlordiazepoxide (Librium), clonazepam (Klonopin), lorazepam (Ativan), oxazepam (Serax), prazepam (Centrax), tamazepam (Restoril), and triazolam (Halcion).

Benzodiazepines stimulate the activity of gamma-aminobutyric acid (GABA), a neurotransmitter in the brain. Perhaps the biggest drawback of these drugs is that they can be addictive. Another serious problem with benzodiazepines is that because they affect brain chemistry, they can cause behavioral and mental problems, including severe memory impairment, confusion, hallucinations, extreme irritation, aggressiveness, and strange behavior. Other side effects associated with the use of benzodiazepines include drowsiness, dizziness, impaired coordination, blurred vision, nausea, indigestion, diarrhea or constipation, headache, and allergic reactions.

Other Natural Remedies for Depression

SAM-e is not the only natural remedy useful for the treatment of depression. St.-John's-wort, 5-HTP, and ginkgo biloba are three of the more common and popular alternative antidepressants on the market. At this time, it appears that SAM-e is more effective than these three options in many cases and overall has fewer side effects. Everyone's body chemistry and needs differ, however, and so it is to your advantage to have choices and, if appropriate, to combine use of SAM-e with another natural antidepressant (see Chapter 9 on how to take SAM-e with other supplements).

St.-John's-Wort
St.-John's-wort (*Hypericum perforatum*) is a popular herbal antidepressant readily available over-the-

counter. The herbal extract of St.-John's-wort has been thoroughly investigated in twenty-five double-blind controlled studies that included a total of 1,592 patients. (Fifteen of the studies compared St.-John's-wort to placebo; ten compared it to an antidepressant—amitriptyline, imipramine, or maprotiline). Patients in all of the studies experienced improvement in depression, anxiety, sleep disturbances, insomnia, and feelings of worthlessness. Overall, St.-John's-wort outperformed the placebo 59 percent to 20 percent, and the antidepressants 64 percent to 58 percent. Although the difference in the benefit between St.-John's-wort and the antidepressants was not great, the herb does not cause the side effects or cost nearly as much as the antidepressants.

St.-John's-wort is not without side effects, although they are far fewer and less severe than those associated with prescription antidepressants. These side effects include loose bowels, sexual dysfunction, nervousness, and in people who have fair skin, photosensitivity. It also can take four weeks or longer to begin offering its antidepressant benefits.

5-HTP

5-HTP is a natural compound that converts to serotonin in the body and also helps increase the levels of other neurotransmitters that are often low in people with depression. Supplements of 5-HTP are extracted from the seed of an African plant called a *Griffonia simplicifolia.*

Many double-blind studies have shown that 5-HTP is just as effective as the selective serotonin reuptake inhibitors such as fluoxetine (Prozac), sertraline (Zoloft), fluvoxamine (Luvox), and peroxatine

(Paxil), and the tricyclics imipramine and desipramine, yet it has much fewer and milder side effects than any of the drugs. For example, in one double-blind study that compared 5-HTP to fluvoxamine, the patients received either 100 mg of 5-HTP or 50 mg of fluvoxamine three times a day for six weeks. A standard test for depression was used to evaluate the degree of depression both before treatment started and when the study was completed. Patients who took 5-HTP showed more improvement in depressed mood, anxiety, physical symptoms, and insomnia than did patients taking fluvoxamine. Fourteen patients (38.9 percent) who took 5-HTP reported side effects, compared with eighteen (54.5 percent) of those taking fluvoxamine. Although these figures are not far apart, the side effects caused by 5-HTP were rated very mild to mild, while those caused by fluvoxamine were rated moderate to severe.

Taking SAM-e with 5-HTP can enhance the work of the latter supplement. Gabriel Cousens, M.D., who is also a psychiatrist, an author, and director of the Tree of Life Rejuvenation Center in Patagonia, Arizona, reports that he and his patients have been very pleased with the results when combining the two natural antidepressants. Overall, his success rate has been more than 90 percent.

Ginkgo Biloba

The extract of the ginkgo biloba leaves has been effective in improving depression in older people. This particular use for the herb came after researchers who gave ginkgo to patients who had experienced cerebrovascular insufficiency (insufficient blood flow to the brain) discovered that the patients had an improvement in mood. This finding prompted several double-blind studies to determine the ability of ginkgo to treat depression. In a 1993 study, forty patients (age range fifty-one to seventy-eight years) with depression who had not improved after taking conventional antidepressant drugs were given either 80 mg of ginkgo biloba three times a day along with their standard antidepressant, or placebo and their antidepressant. After eight weeks, the patients in the ginkgo group had improved significantly on the standard depression test, but those in the placebo group had barely shown any improvement.

This study is important for several reasons. One, it shows that ginkgo can be used to enhance the effectiveness of conventional antidepressants. And because ginkgo is very safe and side effects are uncommon and generally mild (stomach upset, headache, and dizziness), most people find it easy to take. Two, and more importantly, ginkgo appears to be safe to take with another antidepressant: SAM-e. Although it can take up to twelve weeks before ginkgo's benefits become apparent, SAM-e can enhance the herb's effectiveness. Many people take ginkgo for reasons other than depression as well, including senility, premenstrual syndrome, inner ear problems (ringing in the ears, vertigo), and to increase blood flow in the brain.

SAM-e and Depression

The story of SAM-e and its use for treatment of depression began in the early 1970s in Italy, when several researchers learned how to produce SAM-e using cell cultures. A few years later, another Italian research team, which included one of the scientists who had made SAM-e in the laboratory, was conducting a study of the effects of SAM-e on patients with schizophrenia. The investigators noted that although their patients' schizophrenic symptoms were not improving, their depressive mood was. They had inadvertently discovered that SAM-e was a treatment for depression. They reported their findings in an Italian language psychiatric journal, and interest was aroused in the medical community. Now, more than twenty-five years later, dozens of clinical trials conducted around the world have supported the original findings: when it comes to the treatment of depression, SAM-e is superior to placebo and performs equally well, and often better, than certain antidepressant drugs.

SAM-e has the distinction of being the most well-documented nondrug antidepressant on the market today, although it is closely rivaled by St.-John's-wort. It hit the U.S. marketplace at a time when depression affects more than 17 million Americans. This high number of depressed individuals is the main contributor to the high suicide rates in the U.S. Suicide is the fifth leading cause of death among people between the ages of twenty-five and forty-four and the third cause among those aged fifteen to twenty-four. In 1995, the number of suicides in the U.S. was greater than the number of homicides.

SAM-e also comes at a time when people are turning to complementary medicine in increasing numbers. Proof that alternative medical treatments had finally gained a foothold in the mainstream medical community came in 1992 when the National Institutes of Health created the Office of Alternative Medicine. This agency is charged with investigating alternative medicine practices to determine the effectiveness, feasibility, and safety of various complementary treatments, including antioxidant therapy for the prevention of atherosclerosis, and relaxation therapy for caregivers of Alzheimer's disease patients. Thus the environment is ripe for nondrug treatments.

Understanding Depression Studies

Before you read about the studies that demonstrate the effectiveness of SAM-e in the treatment of depression, it helps to understand the standards by which SAM-e, and other antidepressants, are judged. Here are some terms you will encounter:

- **Uncontrolled trial:** A trial in which the subjects' response to a substance is studied over time but is not compared against the response of other subjects who are taking another substance.
- **Controlled trial:** A trial in which the subjects' response to a substance is studied over time and is compared against the response of other subjects who are taking one or more other substances.
- **Placebo:** A substance that resembles the substance being tested except that it is inert (has no effect; a sugar pill).
- **Hamilton Depression Scale (HAM-D):** A standard interview tool used by psychiatrists and psychologists to determine the severity and extent of depression and to measure an antidepressant's effectiveness. The HAM-D consists of a list of symptoms associated with depression. Patients are tested before treatment to establish

their baseline, or starting level, of depression; and after treatment they are tested again to identify the amount of change, whether positive, negative, or no change. Individuals who show a 50 percent or greater decrease in the number of depressive symptoms are called **responders**, and the response is referred to as a **complete response**. If patients show less than a 50 percent decrease in the number of depressive symptoms, the response is referred to as a **partial response**.

- **Double-blind study**: A study in which neither the individuals receiving the tested substance(s) nor the people who are administering them know the true identity of the substances until after the study is completed.

SAM-e and Depression: The Evidence

If you are an individual who likes to know there has been a lot of published, scientific research about a supplement before you decide to take it, SAM-e fits that requirement. Ever since 1978, when SAM-e was first found to be effective in the treatment of depression, dozens of studies have been conducted in patients with depression, and the results have been published in scientific journals. SAM-e joins the list of other natural supplements and herbs that are being marketed for depression, including St.-John's-wort, ginkgo biloba, and 5-HTP. Although these other natural substances have been studied for their ability to elevate mood and have been found to be effective for many people, the claims made about these products are not supported by the same wealth of scientific evidence that surrounds SAM-e.

Overall, the studies of SAM-e's effects on depression show that:

- SAM-e relieves depression and depressive symptoms equally as well or better than conventional antidepressants.
- SAM-e provides relief from depression and depressive symptoms at least twice as fast as conventional antidepressants.
- SAM-e causes virtually no side effects, except for mild nausea or headache in some cases.
- SAM-e does not cause withdrawal symptoms when patients stop taking it.
- SAM-e is nontoxic, based on studies conducted thus far, even when taken at doses higher than recommended.

Here's a closer look at the studies that have provided this information.

Meta-Analyses

In an effort to better understand and appreciate the information contained in the many clinical trials on SAM-e, doctors have conducted meta-analyses of the clinical studies that included the use of SAM-e in the treatment of depression. A meta-analysis is a study in which many smaller studies with similar characteristics are combined and evaluated as if they were one study. This process allows investigators to extract a broader scope of information than can be obtained from the many individual studies.

In 1988, P. G. Janicak and his associates of the Psychiatric Institute, University of Illinois, in Chicago, performed a meta-analysis of the studies that had been conducted with SAM-e up to the year 1987.

All the studies involved the use of injectable SAM-e. (Since that time, more recent studies have demonstrated that oral SAM-e is just as effective as an intravenous dose. However, a higher dose of the oral preparation is needed to compensate for a difference in how SAM-e is absorbed orally versus through an injection.) The Janicak analysis looked at several different groups of studies. One group included seven double-blind controlled trials in which SAM-e had been compared with placebo. Dr. Janicak's team found that 78 percent of patients who took SAM-e were responders (29 of 37 patients) compared with 4 percent of patients (1 of 25 patients) who took a placebo.

The Janicek team also evaluated the results of nine controlled clinical studies in which SAM-e had been compared with various antidepressants, including imipramine, amitriptyline, clomipramine, and nomifensine. Among people who were given SAM-e, 109 of 142 people were responders, compared with 80 of 124 responders among those who were taking one of the mentioned antidepressants. The final score: 76 percent of people who took SAM-e got relief, compared with 61 percent of people who took conventional antidepressants.

In 1994, the results of a meta-analysis conducted by G. M. Bressa, of Rome, Italy, were published. Dr. Bressa reviewed thirty-eight research projects that involved the use of SAM-e in depression: thirteen uncontrolled trials and twenty-five controlled studies. The uncontrolled trials had a total of 377 participants, who had been studied for their response to SAM-e alone. Some participants had received SAM-e by injection (either into the muscle or intravenously) and others had received an oral dose of 1,600 mg daily. Overall, the patients' response to SAM-e was positive, and the findings helped support the positive results of controlled trials. However, because uncontrolled trials do not include a comparison group (control group), they carry limited scientific weight.

Dr. Bressa also evaluated twenty-five controlled studies that involved a total of 793 patients. Six of the studies involved a comparison of SAM-e (1,600 mg given orally or a comparative amount given via injection) with placebo. The response rate in five of the studies was greater for SAM-e than it was for placebo, but it was not much different in the sixth study. When all the responses were considered together, SAM-e produced a 70 percent partial response rate, compared with a 30 percent partial response rate for placebo. When just the complete responses were considered, the percentages were closer together: 38 percent of patients who took SAM-e had a complete response compared with 22 percent of patients who took placebo.

The remaining nineteen controlled studies analyzed by Dr. Bressa involved a comparison of SAM-e with various tricyclic antidepressants: amitriptyline, chlorimipramine, desipramine, and imipramine. Patients given SAM-e received either 1,600 mg orally or a comparable amount via injection. Overall, the partial response rates were 92 percent for SAM-e and 85 percent for the antidepressants. The complete response rates were 61 percent for SAM-e and 59 percent for the antidepressants. Although SAM-e did not show a big advantage over the tricyclics in terms of response, it did not cause the bothersome side effects associated with the antidepressants. This feature is considered to be of vital importance, because when patients don't experience side effects, they continue to take their antidepressants.

Controlled Studies

In 1994, the results of a study conducted at the University of California, Irvine, were published. The researchers compared the response of twenty-six patients with major depression to either SAM-e or the tricyclic antidepressant desipramine. At the end of the four-week, double-blind study, 62 percent of the patients who had received SAM-e had improved, compared with only 50 percent of the patients who had taken desipramine. The researchers in this study also had another finding. Regardless of whether patients took SAM-e or desipramine, all of the patients considered to be responders also had a significant increase in their plasma concentration of SAM-e. This finding indicates that one way tricyclic antidepressants work is to raise SAM-e levels in the body, and offers further support for the use of SAM-e in treating depression.

Another study, conducted by M. Fava and associates in 1995, is notable because it illustrates how quickly SAM-e's antidepressive benefits can take effect. While conventional antidepressants take a minimum of twenty-one days to take effect, the 195 patients in Dr. Fava's study responded to SAM-e in seven to fourteen days. A rapid response to treatment is critical in depressed patients, because of the potential for overdosing. Studies have shown that many patients taking conventional antidepressants overdose on those drugs during the "waiting period" between starting the medication and getting relief, a time that can seem like an eternity for depressed individuals, especially when they must also deal with the side effects of those drugs at the same time. SAM-e can cut the waiting time by up to two-thirds and eliminates the adverse reactions.

SAM-e as Helper/Enhancer

An interesting study performed by C. Berlanga and associates in 1992 revealed the ability of SAM-e to enhance the action of conventional antidepressants. Dr. Berlanga's team conducted a double-blind clinical trial that included forty patients who had moderate to severe depression. Half the patients received SAM-e and the tricyclic antidepressant imipramine, while the other half received placebo and imipramine. The researchers found that the patients who were given the SAM-e/imipramine combination got relief from their depression sooner than the patients who were taking imipramine and placebo.

The use of SAM-e as a treatment enhancement is common in Europe. That fact, along with the results of Berlanga's study, prompted Sol Grazi, M.D., author of *European Arthritis and Depression Breakthrough! SAMe,* to say that "At the very least, it's worth considering using SAMe to 'jump start' conventional antidepressants by combining both during the first two to three weeks of therapy." See Chapter 9 for more on how to take SAM-e.

SAM-e and Depression Caused by Preexisting Medical Conditions

SAM-e has been found to relieve the depression that accompanies many preexisting medical conditions. A research team in Italy studied the effects of SAM-e on depressive symptoms caused by different medical conditions in a group of fifty-five patients. All of the patients had moderate to major depression, as well as one of the following medical conditions: liver disease, insulin-dependent diabetes, hepatitis, viral pneumonia, psoriasis, cardiovascular disease, cancer, obesity, bronchial asthma, endocrine diseases, herniated disk, and congenital hip dislocation. Forty of the patients were hospitalized and received injections of SAM-e. The fifteen outpatients took oral doses of SAM-e (two 400-mg tablets daily) for four weeks. Depression improved significantly in all patients, and none of them experienced side effects that led them to withdraw from the study.

This is welcome news for several reasons. One is the high prevalence of depression in people with physical disease. According to an article in a 1996 issue of the *Journal of the American Medical Association*, depressive disorders occur in 50 percent or more of people who have medical illnesses. Among stroke patients, for example, depression is a problem in about 50 percent of patients with a stroke in the left side of the brain and in 10 percent of those with a right-sided stroke, according to Stuart Yudofsky, M.D., a professor at Baylor College of Medicine, Houston, Texas. People with cancer also have a high prevalence of depression. J. Stephen McDaniel, M.D., associate professor of psychiatry at Emory University School of Medicine in Atlanta, Georgia, notes that the mean prevalence of depression among cancer patients is 24 percent, but that the rate varies depending on the type of cancer. Among people with pancreatic cancer, for example, the rate is 50 percent. A lifetime prevalence rate for depression among people with type II diabetes is 25 percent, says Paul Goodnick, M.D., professor of psychiatry at the University of Miami School of Medicine.

Thus, depression among people with medical conditions is a significant problem, and one that is not always recognized by physicians who treat these patients. Failure to identify and treat depression in these individuals can make their medical conditions worse. For example, it may cause people to not take their medications or to not comply with other treatments or lifestyle changes recommended or essential for management of their disease. A diabetic who is depressed may eat sweets and junk food and experience episodes of hyperglycemia (high blood sugar). A person who has suffered a stroke may not do the necessary physical therapy that can get her back on her feet.

The ability to use SAM-e for depression in people who also have a medical condition is especially important for patients who cannot take conventional antidepressants because of their illness. One group of individuals who fit this description are those with myocardial infarction. According to Steven Roose, M.D., professor of psychiatry at New York Psychiatric Institute, 20 percent of patients who have suffered a myocardial infarction have major depression. The death rate among these patients is greater than that of their counterparts who are not depressed. Treatment with tricyclic antidepressants may not be safe for these individuals, however, because some tricyclics have antiarrhythmic properties and may increase the risk of death in these patients. Use of SAM-e in these patients and others in similar situations could be a natural, safe choice.

SAM-e and Depression in the Elderly

Late-life depression affects an estimated 6 million Americans, but only about 10 percent of these individuals ever receive treatment. Among the elderly who do receive treatment for depression, several complications commonly occur. One is the incorrect administration or use of drugs. According to a recent study published in *Psychosomatics*, 64 percent of 131 elderly patients who were being admitted to a geriatric psychiatric unit had been given benzodiazepines inappropriately. Another study shows that when primary care physicians do prescribe antidepressant drugs to elderly patients, often the dose, the duration of treatment, or both are insufficient.

Another problem is that many elderly individuals take several different medications for various physical ailments. This situation, known as polypharmacy, can be serious, and even deadly, especially given the fact that the elderly have an increased sensitivity to the effects of drugs. With polypharmacy, the chances of toxicity and delirium in the elderly increase, and taking any conventional antidepressant has the potential to worsen delirium.

SAM-e does not appear to cause any of these problems in the elderly. It also seems to begin working on depressive symptoms much sooner than conventional antidepressants. The elderly have a slower metabolism than younger people and so antidepressants typically take at least four to six weeks to show any benefit. SAM-e begins in half that time or less. (Information on how to take SAM-e is in Chapter 9.) Naturally, no one should begin any type of antidepressant treatment without first consulting with their physician and having their current health status and medication routine reviewed.

What the Doctor Ordered?

During a recent interview with Gabriel Cousens, M.D., a general practitioner, psychiatrist, and family therapist practicing in Patagonia, Arizona, he noted how he was using SAM-e to treat between forty and fifty individuals with depression. Dr. Cousens reported a success rate of better than 90 percent and said only one patient had experienced very mild and temporary stomach distress. He and other doctors across the country are singing the praises of SAM-e. Although depression relief is not 100 percent guaranteed (it's not with any drug or supplement), many studies have shown SAM-e to be more effective and better than its alternatives:

- Antidepressant drugs have complete and partial response rates that are below those obtained with SAM-e.
- Discomforting and often debilitating side effects are generally associated with conventional antidepressant drugs.
- It takes at least three weeks, and sometimes longer, for conventional antidepressants to begin their antidepressant benefits.
- The necessity of getting a prescription, even if you have only mild depression, is inconvenient, plus most people would prefer to keep their treatment a private matter.

SAM-e offers a 70 percent chance of response; quick results; minimal chance of side effects; and the convenience of buying it at your corner drugstore, health food store, or other retail outlet, or via mail order or the Internet.

It could be just what the doctor ordered.

NOTES

Agnoli, A., Andreoli, V., Casacchia, M., et al. "Effect of S-adenosyl-methionine upon depressive symptoms." *J Psychiatr Res* 13 (1976): 43–54.

Bell, K. M., Potkin, S. G., Carreon, D., and Plon, L. "S-adenosyl-methionine blood levels in major depression. Changes with drug treatment." *Acta Neurol Scand Supp* 154 (1994): 15–18.

Berlanga, C., Ortega-Soto, H. A., Ontiveros, M., and Senties, H. "Efficacy of S-adenosyl-L-methionine in speeding the onset of action of imipramine." *Tlalpan Psychiatry Res* 44 (3) (1992): 257–262.

Brown University Long Term Care Quarterly, 1997.

Carney, M. W. P., Chary, T. K. N., Bottiglieri, T., et al. "Switch mechanism in affective illness and oral S-adenosylmethionine (SAM)." *Br J Psychiatry* 150 (1987): 724–725.

Carney, M. W. P., et al. "Neuropharmacology of S-adenosylmethionine." *Clin Neuropharmacol* 9(3) (1986): 23–24.

DeLeo, D. "S-adenosylmethionine as an antidepressant: A double-blind trial versus placebo." *Curr Ther Res* 7 (1987): 254–257.

Fazio, C., Andreoli, V., Agnoli, A., et al. "Therapeutic effects and mechanisms of action of S-adenosylmethionine (SAM) in depressive syndromes." [in Italian] *Minerva Med* 64(20) (30 April 1997): 1515–1529.

Janicak, P. G., Lipinski, J. D., Comaty, J. E., et al. "S-adenosylmethionine. A literature review and preliminary report." *Alabama J Med Sci* 25 (1988): 306–313.

Janicak, P. G., et al. "Parenteral S-adenosylmethionine in depression: A literature review and preliminary report." *Psychopharmacology Bull* 25 (1989): 238–241.

Kufferle, B., and Grunberger, J. "Early clinical double-blind study with S-adenosyl-L-methionine: A new potential antidepressant." In Costa, E., and Racagni, G., eds., pp. 175–180. *Typical and Atypical Antidepressants: Clinical Practice.* New York: Raven Press, 1982.

Lamberg, L. "Treating depression in medical conditions may improve quality of life." *Journal of the American Medical Association*, 18 September 1996.

Lipinski, J. F., Cohen, B. M., et al. "Open trial of S-adenosylmethionine for treatment of depression." *Am J Psychiatry* 141 (March 1984): 448–450.

Mantero, P., Pastorino, P., Carolei, A., et al. "Controlled double-blind study (SAM-e imipramine) in depressive syndromes." [in Italian] *Minerva Med* 66 (1975).

Miccoli, L., Porro, V., and Bertolino, A. "Comparison between the antidepressant activity of S-adenosyl-L-methionine (SAMe) and that of some tricyclic drugs." *Acta Neurologica* 33 (1978): 243–255.

Murray, Michael T. *Natural Alternatives to Prozac.* New York: William Morrow & Co., 1996.

Muscettola, G., Galzenati, M., Balbi, A. "SAMe versus placebo. A double-blind comparison in major depressive disorders." In Costa, E., and Racagni, G., eds., pp. 151–156. *Typical and Atypical Antidepressants: Clinical Practice.* New York: Raven Press, 1982.

Orrell, M., et al. "Management of depression in the elderly by general practitioners. Use of antidepressants." *Family Practice* 12(1) (1995): 5–11.

Scaggion, G., Baldan, L., and Domanin, S. "Antidepressive action of S-adenosylmethionine compared to nomifensine maleate." [in Italian] *Minerva Psichiatr* 23 (1982): 93–97.

Scarzella, R., and Appiotti, A. "A double-blind clinical comparison of SAMe vs cloipramine in depressive disorders." [in Italian] *Riv Sper Freniatr* 102 (1978): 359–365.

Schubert, H., and Halama, P. "Depressive episode primarily unresponsive to therapy in elderly patients: Efficacy of *Ginko biloba* in combination with antidepressants." *Geriatr Forsch* 3 (1993): 45–53.

Vorberg, G. *Clinical Trials Journal* 22 (1985): 149–157.

5

Pain, Pain, Go Away: SAM-e and Osteoarthritis

It may be hard to identify which group is more enthusiastic about SAM-e: those who hail it for its antidepressive powers, or those who tout it for its ability to alleviate the pain and stiffness associated with osteoarthritis. Overall, in sheer number of patients studied, many more have participated in studies of SAM-e and osteoarthritis than SAM-e and depression. And the benefits from SAM-e in people who suffer with osteoarthritis have been impressive, as you will read in this chapter. This is good news to the more than 40 million people in the United States who have osteoarthritis, but it is old news to the people in European countries. They have been using SAM-e to treat symptoms of osteoarthritis for more than twenty years. In Germany, it is an approved treatment for this rheumatic disease.

If you or someone you know has osteoarthritis, you'll want to read this chapter. It explains how osteoarthritis affects the body, what causes it, and the pitfalls of the current conventional treatments for this disease—problems you may be able to avoid if you take SAM-e. It also discusses some of the natural treatments of osteoarthritis, including glucosamine, chondroitin, and MSM.

What Is Osteoarthritis?

Of the more than one hundred different types of arthritis, osteoarthritis is the most common. All types of arthritis share a common characteristic, reflected in the meaning of the word "arthritis": inflammation of the joints. Osteoarthritis ("osteo" means bone) is also known as *degenerative joint disease*, because it involves the degeneration of the joint and the loss of cartilage, which is a gel-like material found between the joints where the bones meet. As the cartilage wears down, it causes friction between the bones. Over time, the structures that hold the joints together—the ligaments, muscles, and tendons—become weaker, and the joints become painful, stiff, and deformed.

Although osteoarthritis is not strictly a disease of older people, 80 percent of men and women older than fifty have the disease. According to the National Institute of Arthritis and Musculoskeletal Diseases, osteoarthritis is the number one cause of disability for people older than sixty-five, even more than back pain, heart and lung problems, cancer, and diabetes. And as a large number of baby boomers begin to enter their sixth decade, the number of people affected with osteoarthritis will continue to rise.

Joints in the hands and those that bear weight—feet, knees, spine, and hips—are most often affected by osteoarthritis. Symptoms of osteoarthritis include early morning stiffness in the joints, stiffness following periods of rest or inactivity, pain that gets worse with movement, and tenderness around the affected joints. As the joints deteriorate, many people develop bony swellings and nodes, especially in the hands, which can severely limit mobility and joint function. Osteoarthritis of the hips causes localized pain and a limp, while osteoarthritis of the spine can cause the nerves and blood vessels to compress, resulting in pain and restricted blood flow.

Osteoarthritis can be primary or secondary. Primary osteoarthritis is considered to be the degeneration of the cartilage, beginning around age fifty, in people who have no obvious abnormalities that would cause the wear-and-tear process. This type of osteoarthritis is usually attributed to the cumulative effect of normal use, as well as contributing factors such as being overweight or having done a great deal of repetitive movements, such as assembly line work. Secondary osteoarthritis is caused by a predisposing factor, such as preexisting abnormal cartilage, trauma (e.g., fractures, surgery), or the presence of a previous inflammatory joint disease, such as gout or rheumatoid arthritis.

Causes of Osteoarthritis

Researchers have not identified the exact cause, or causes, of osteoarthritis, although they do know that stress on the joints, caused primarily by repetitive movement, obesity, and physical trauma such as injuries, is a major factor. Other factors may include heredity, fluctuations in the body's biochemistry, changes in hormone production, or inflammatory disease.

To understand why SAM-e is an effective treatment for osteoarthritis, it helps to know how osteoarthritis evolves. The cartilage and its components are at the center of the degeneration process. Cartilage consists of three main factors: chondrocytes, collagen, and proteoglycans. Chondrocytes are cells which produce chondroitin sulfate, collagen, and proteoglycans. If chondroitin sulfate sounds familiar, it's probably because it is sold as a natural supplement to treat osteoarthritis. (Chondroitin sulfate is discussed later in this chapter.) Collagen is a protein that forms a network of connective tissue that supports the cartilage. Proteoglycans are large molecules found on the outside of the collagen. They perform a critical task in helping keep joints healthy because they attract and hold onto water. In this way they have an important role in maintaining the synovial fluid, the clear lubricating liquid that acts as a shock absorber for the joints and helps keep them moving smoothly.

In healthy individuals, the amount of proteoglycan that is produced and broken down remains relatively stable, which helps ensure healthy joints and good mobility. But when the amount of stress placed on the joints becomes too great, the amount of proteoglycan and collagen that is destroyed becomes greater than the amount produced. Stress also limits the ability of the chondrocytes to manufacture more proteoglycans and collagen. Without a sufficient amount of materials to build and repair cartilage, it begins to deteriorate. This is the start of osteoarthritis.

Fortunately, chondrocytes have the ability to repair damaged cartilage, but they need an energy supply to do that task. Here is where a catch-22 situation arises: stress causes damage to the cartilage, but the chondrocytes can't repair the cartilage unless it gets glucose (energy). And a major source of that energy is—you guessed it—the cartilage. Thus osteoarthritis develops when the supply of chondrocytes is insufficient, which means they cannot produce enough proteoglycans and collagen, and they cannot do repairs. Once scientists understood this process, they realized that one way to help prevent or treat osteoarthritis is to ensure a healthy level of chondrocytes. There is only one problem: the drugs that are being used to treat osteoarthritis are not doing that; in fact, they are doing the opposite. (See "The 'Other' Big Problem with NSAIDs" below.)

Conventional Treatments for Osteoarthritis

The standard treatments for osteoarthritis include aspirin and other nonsteroidal anti-inflammatory drugs (NSAIDs). The most common NSAIDs are listed in the sidebar. Nonsteroidal anti-inflammatory drugs relieve pain mainly by reducing inflammation, which they do by suppressing the action of substances called prostaglandins. Prostaglandins are hormones that can cause inflammation and increase the awareness of pain.

The Problem with NSAIDs
The use of NSAIDs in osteoarthritis, although very widespread, can create several problems. One is that they can cause gastrointestinal discomfort, ulcers, stomach bleeding, dizziness, headache, and ringing in the ears. Individuals who have heart failure, kidney

NSAIDS

Common NSAIDs on the market: Aspirin, choline magnesium trisalicylate, diclofenac (Voltaren), diflunisal (Dolobid), fenoprofen (Nalfon), flurbiprofen (Ansaid), ibuprofen (Motril, Advil), indomethacin (Indocin), ketoprofen (Orudis, Oruvail), meclofenamate (Meclomen), nabumetone (Relafen), naproxen (Aleve, Naprosyn), oxaprozin (Daypro), phenylbutazone (Azolid, Butazolidin), piroxicam (Feldene), salsalate (Disalcid), sulindac (Clinoril), tolmetin (Tolectin)

disease, or cirrhosis should consult with their physician before taking NSAIDs, because they can cause kidney toxicity and result in death.

Some people mistakenly believe that if their aspirin or NSAID is enteric-coated it protects them against experiencing stomach problems. But a coating of calcium carbonate, magnesium oxide, or other antacid does not prevent stomach damage from occurring.

Another class of pain-relieving drugs, called COX-2 inhibitors, was introduced to the market in December 1998. The first to become available to the public was Celebrex (celecoxib), which the FDA approved for the treatment of osteoarthritis and rheumatoid arthritis only (that is, it was not approved for pain management of any other condition). Other COX-2 inhibitors are expected to become available soon, including Vioxx (rofecoxib). Although one of the attractions of COX-2 inhibitors is supposed to be their lack of gastrointestinal side effects, they do cause gastroduodenal ulcer in 7 percent of people who take them.

Celecoxib also causes side effects similar to those associated with NSAIDs, including hypertension and swelling, and should not be taken by anyone who has heart failure, liver disease, preexisting asthma, or water retention, or by anyone who is taking lithium or fluconazole, without first consulting with a physician.

The "Other" Big Problem with NSAIDs

It's been suspected for many years that aspirin and certain other drugs used to relieve the symptoms of osteoarthritis actually contribute to the disease by inhibiting the ability of the chondrocytes to repair the cartilage. If you're thinking, "Wait, you mean the drugs I'm taking actually cause more harm than good? The drugs that are supposed to relieve pain are actually worsening the condition that causes the pain?" you're right.

In 1987, Dr. K. Brandt and his research team reported on this phenomenon in an article in the *American Journal of Medicine*. In it they stated that aspirin and several other NSAIDs had a negative effect on proteoglycan metabolism. They noted that in studies of animals with osteoarthritis, aspirin appeared to speed up cartilage damage and that several NSAIDs inhibited the ability of chondrocytes to produce proteoglycans. In conclusion they warned that if these observations in animals were also true in patients with osteoarthritis who were taking NSAIDs, cartilage destruction could be accelerated.

Brandt was not alone in his suspicions, and many other studies followed. Researchers investigated the effects of aspirin, etrodolac, tenoxicam, and other NSAIDs and found that these drugs help destroy cartilage in two ways: by inhibiting the ability of the chondrocytes to produce proteoglycan, and by preventing the production of chondroitin sulfate. Chondroitin sulfate is a major component of the connective tissue. One of its primary functions is to attract water to the cartilage, which helps prevent breakdown of the cartilage and promotes joint mobility.

In 1995, D. H. Manicourt and his research team reported more bad news: that the damaging effects of NSAIDs were actually worse in people with osteoarthritis than in those with normal cartilage. They suggested that this occurred because damaged cartilage attracts more of the NSAIDs than does healthy cartilage. Thus the people who need the most protection and who think they are getting it are actually getting the least help and exposing themselves to the most damage.

Now for some good news. SAM-e not only doesn't worsen the wear-and-tear process on the joints; it is believed to actually repair the damaged cartilage. Here's the proof.

SAM-e and Osteoarthritis

In 1987, the *American Journal of Medicine* dedicated its entire 20 November issue to SAM-e and osteoarthritis. In the issue were a review of the clinical studies by C. Padova as well as the published findings of the individual clinical studies themselves. The first thing that is impressive about the studies is the sheer number of patients who were treated with SAM-e: 21,524 over a period of five years. Dr. Padova summed up the wealth of information by saying, "The intensity of therapeutic activity of SAMe against osteoarthritis is similar to that exerted by nonsteroidal anti-inflammatory drugs, but its tolerability is higher. Based on these findings, SAMe is proposed as the prototype of a new class of safe drugs for the treatment of osteoarthritis."

First let's take a look at some of the studies included in this review. (See Chapter 4 for a list of terms used to describe studies.) When determining the effectiveness of a particular treatment for osteoarthritis, investigators use a combination of reports from the patients on the severity of pain and what they observe of the patients. In all of the studies in this chapter, the criteria usually measured included: amount of morning stiffness, pain at rest, pain when moving, amount of swelling at the affected joints, amount of limited motion in affected joints, and the amount of noise joints made when moved. Generally, people who take SAM-e for osteoarthritis experience some relief after about two weeks of treatment.

SAM-e in Laboratory Studies

Before drugs and other substances are given to human subjects, years of testing are done on laboratory animals. One study, conducted by Hector A. Barcelo, M.D., and his research staff, looked at the specific action of SAM-e in the cartilage. They found that administration of SAM-e to rabbits increased the "numbers of cells per square millimeter, formation of cellular clones, and maintenance of cartilage depth." SAM-e also was associated with "superior retention of proteoglycans," which, as mentioned above, are necessary for healthy joints. Based on the results of their study, Dr. Barcelo and his team rated SAM-e as a good protector of chondrocytes and noted that it also had painkilling and anti-inflammatory effects.

Dr. M. F. Harmand and his colleagues in France conducted a study in which they explored the effect of SAM-e on chondrocyte cultures obtained from human osteoarthritic bone. "Treatment with SAM-e was associated with decreased cartilage loss, superior retention of proteoglycans . . . and evidence of stimulated chondrocyte proliferation," noted Dr. Roland W. Moskowitz, who reviewed their study in the *American Journal of Medicine*. Addition of SAM-e to the cultures resulted in an increase in protein activity, especially for the proteoglycans.

Both of these studies provide evidence of SAM-e's ability to restore cartilage in osteoarthritis, a major benefit in treatment.

SAM-e in Uncontrolled Trials

The largest and most ambitious trial was conducted by Drs. R. Berger and H. Nowak of Germany. Their uncontrolled trial included 20,641 patients with osteoarthritis of the spine, hip, fingers, and knee. All of the patients were given oral doses of SAM-e for a total of eight weeks. The patients were not allowed to take any other medications for pain or arthritis, but they were permitted to continue taking medications for any other medical conditions they had. When asked about how effective SAM-e was for relief of osteoarthritis symptoms, 71 percent described it as "very good" or "good," 21 percent said "moderate," and 9 percent said "poor." When tolerance (number and severity of side effects) was evaluated, 87 percent rated it as "very good" or "good," 8 percent as "moderate," and 5 percent as "poor."

The second largest trial was carried out by B. Konig, of the University of Mainz, Germany. This was a multicenter trial, which means patients from different treatment sites were included in the study. A total of 97 patients who had osteoarthritis of the spine, knee, or hip finished the two-year study. During the first two weeks of the study, the patients received 600 mg of SAM-e daily and then received 400 mg daily for the remainder of the study. The patients' symptoms were evaluated at the end of the first and second weeks of treatment and then monthly for the remainder of the study. Patients noticed improvement after only one week of treatment, and the researchers reported that "the major portion of therapeutic success was already obtained after four weeks of therapy." This relief continued up until the end of the study. Twenty patients reported mild side effects, but none of the patients dropped out of the study because of them. As a side benefit, patients who had been suffering with depression reported an improvement in their depressive feelings. Dr. Konig concluded that "From these data, we suggest that SAM-e is an innovative alternative to the drugs currently used in the treatment of patients with osteoarthritis."

SAM-e in Controlled Trials
In a controlled trial conducted in Germany by Dr. H. Muller-Fassbender, the effectiveness of SAM-e was compared with that of ibuprofen in thirty-six patients who had osteoarthritis of the hip, knee, and/or spine. This was a double-blind study in which treated patients received either 1,200 mg of SAM-e daily or 1,200 mg of ibuprofen daily for four weeks. All of the patients were evaluated before and after treatment for

pain symptoms. At the end of the trial, both groups had similar improvement and had tolerated the treatments well. Results like these are important because people who cannot take ibuprofen may be able to get equally effective results by taking SAM-e, and not experience any of the possible side effects associated with the ibuprofen.

The effectiveness of SAM-e was compared with that of indomethacin in another double-blind trial that included thirty-six patients with osteoarthritis of the knee, hip, and/or spine. During the four-week trial, half the patients were given 1,200 mg of SAM-e daily in an oral dose and the other half received 150 mg of indomethacin daily. Patients' symptoms were evaluated at the beginning and the end of the trial. By the end of week four, both groups of patients had similar, significant improvement in symptoms. Two patients who had taken SAM-e experienced slight nausea during the second week of treatment, while seven patients taking indomethacin had adverse reactions.

In yet another drug comparison trial, a 1,200-mg daily dose of SAM-e was compared with placebo and 750 mg of the NSAID naproxen in a study conducted in Italy. The double-blind study included 734 patients from 33 different medical centers (such a study is typically called a "multicenter trial"). Most of the patients (582) had either hip or knee osteoarthritis. Both SAM-e and naproxen were equally effective in relieving pain, but SAM-e was significantly better tolerated than the drug. The researchers concluded that SAM-e "deserves to be ranked among the most adequate drugs for the medical management of osteoarthritis," based both on its ability to relieve pain and its lack of significant side effects.

A few years before the *American Journal of Medicine* special issue on SAM-e and osteoarthritis, other studies on this subject had been published in several other medical journals. One of the studies looked at the effectiveness of SAM-e versus ibuprofen in 150 patients who had hip and/or knee osteoarthritis. The 1985 study was a multicenter, double-blind trial in which patients received either SAM-e (oral) or ibuprofen, 400 mg three times a day for thirty days. The researchers reported that patients who had taken SAM-e had slightly more improvement in pain relief than those who had taken ibuprofen. Only five patients in the SAM-e group experienced mild side effects, while sixteen patients in the ibuprofen group had adverse effects.

Subsequent Studies

Since the extensive review in 1987, there have been several other studies of the use of SAM-e in osteoarthritis. One, conducted at the Indiana University School of Medicine in 1994 by Dr. J. D. Bradley and his research associates, evaluated SAM-e's effectiveness in relieving symptoms of osteoarthritis of the knee and the speed at which these benefits occur. This study lasted twenty-eight days and involved the use of both intravenous (IV) and oral doses of SAM-e: 400-mg IV given for the first five days of the study, followed by 200-mg tablets given three times a day for twenty-three days. This was a double-blind, placebo-controlled trial, so half of the eighty-one patients in the study received placebo in place of SAM-e.

The study was conducted at two different sites: one site enrolled patients with milder osteoarthritis and the other had more severe cases. The researchers found that among the patients with milder osteoarthritis, those who received SAM-e had a significantly greater reduction in overall pain and pain when resting than the patients who had received placebo. Benefits of SAM-e were experienced as early as fourteen days after starting treatment. Among the patients with more severe osteoarthritis, the response between the patients receiving SAM-e and those getting placebo did not differ.

The researchers concluded that SAM-e "may be an effective treatment for some patients with symptomatic knee osteoarthritis and merits further study." They could not draw any conclusions about the effectiveness of giving IV SAM-e before starting oral treatment.

Glucosamine and Other Natural Remedies

In the treatment of osteoarthritis, SAM-e can be referred to as a "chondroprotective substance," a term used by Dr. H. G. Fassbender of the World Health Organization Center in Mainz, Germany, in his 1987 article "Role of chondrocytes in the development of osteoarthritis." A chondroprotective substance is an agent that protects the chondrocytes and allows them to synthesize proteoglycan and collagen—and thus prevent damage to the cartilage—while providing relief from pain and inflammation. This is the type of remedy needed for the treatment and prevention of osteoarthritis. With SAM-e, relief usually begins after about two weeks of treatment. (See Chapter 9 for how to take SAM-e for osteoarthritis.)

Another natural substance that qualifies as a chondroprotective agent is glucosamine sulfate. This section takes a brief look at glucosamine sulfate and two other natural treatments for osteoarthritis that, while not chondroprotective in action, do provide relief and have virtually no side effects.

Glucosamine and Chondroitin

Ever since the publication of *The Arthritis Cure* by Jason Theodosakis, M.D., B. Adderly, and B. Fox, there has been a tremendous amount of interest in and use of glucosamine sulfate for relief of osteoarthritis. Glucosamine is a compound that occurs naturally in all body tissues. It has a critical role in maintaining the integrity of tissue and is one of the substances that helps cushion the joints. Glucosamine supports the production of proteoglycan and helps increase the amount of chondroitin sulfate (another substance produced by the body) where it is needed most: where cartilage is being repaired. Although glucosamine is not a painkiller, the actions that it performs results in pain relief and reduction of inflammation.

Since the 1980s, clinical studies of glucosamine have shown it to be significantly superior to placebo for reducing pain and joint tenderness and increasing mobility. In studies comparing it with NSAIDs like ibuprofen, it has proved to be as effective in relieving symptoms of osteoarthritis and is much less likely to cause adverse effects. In one double-blind, placebo-controlled trial, for example, 37 percent of patients experienced side effects from ibuprofen but only 7 percent had any side effects from glucosamine. Side effects of glucosamine include mild indigestion, diarrhea, nausea, or heartburn, which are often avoided if it is taken with meals.

Supplements containing both glucosamine and chondroitin sulfate, another chondroprotective agent and a substance that is produced by the chondrocytes, is advocated by some as a treatment for osteoarthritis, while others believe the chondroitin component adds little or nothing to the remedy because the molecules are too large to be absorbed properly. Further studies of the benefits of the glucosamine-chondroitin combination are needed. Information on how to take glucosamine and chondroitin is in Chapter 9.

As for which of the three chondroprotective substances—glucosamine sulfate, chondroitin sulfate, or SAM-e—is more effective in the treatment of osteoarthritis, Dr. Grazi summed it up by saying that "considering the evidence . . . SAMe demonstrat[es] more overall effectiveness at this point than either glucosamine or chondroitin."

MSM

MSM, or methylsulfonylmethane, is another safe, natural substance found in the body and in sulfur-containing amino acids. MSM is involved in many different processes in the body, including the creation of new cells and the production of collagen. Although it is not a chondroprotective substance, it has gained some notoriety as an effective treatment for the pain and tenderness of osteoarthritis. Part of MSM's ability to relieve arthritis pain is attributed to its ability to promote the flow of toxins out of the cells and the passage of nutrients into the cells. This process, which balances the pressure on both sides of the cell walls, helps prevent inflammation in the joints. Both SAM-e and MSM can complement each other for the treatment of osteoarthritis. Information on how to take MSM is provided in Chapter 9.

Stinging Nettles

The effectiveness of this weedy plant in the treatment of osteoarthritis has been investigated by scientists at the University of Frankfurt, the University of Dusseldorf, and St. Elisabeth's Clinic in Straubing, Germany. They conducted a controlled trial in which forty people with severe arthritis were given either 200 mg of the NSAID diclofenac or 50 mg of diclofenac and 50 g of nettle leaves. The patients in both groups had a significant reduction in pain and stiffness, even though the one group took only 25 percent of the diclofenac dosage. This study supports the findings of an earlier trial in which people with arthritis were able to reduce their NSAID dose by half when they added nettle leaves to their treatment plan. Although additional studies are needed to determine if nettle leaves alone are effective in the treatment of osteoarthritis, another possible combination treatment for osteoarthritis is SAM-e and nettles.

Decades of clinical research on SAM-e and osteoarthritis and its practical use among tens of thousands of Europeans all support its use as an effective natural treatment for this painful and often debilitating disease. SAM-e's chondroprotective abilities make it a truly exciting and much-needed remedy, one that will hopefully bring relief to millions of people who until now have had few options from the damaging effects of conventional osteoarthritis medications.

NOTES

Adams, M. E. "Cartilage research and treatment of osteoathritis." *Current Opinion on Rheumatology* 4(4) (August 1992): 552–559.

Barcelo, Hector A., Wiemeyer, Juan, C.M., Sagasta, L., Carlos, et al. "Effect of S-adenosylmethionine on experimental osteoarthritis in rabbits." *American Journal of Medicine* 83(Suppl 5A) (20 November 1987).

Bassleer, C. T., Henrotin Y. E., Regnister J. L., and Franchimont, P. P. "Effects of tiaprofenic acid and acetylsalicylic acid on human articular chondrocytes in 3-dimensional culture." *Journal of Rheumatology* 19(9) (September 1992): 1433–1438.

Bradley, J. D., Flusser, D., Katz, B. P., et al. "A randomized, double blind, placebo controlled trial of intravenous loading with S-adenosylmethionine (SAM) followed by oral SAM therapy in patients with knee osteoarthritis." *Journal of Rheumatology* 21(5) (May 1994): 905–911.

Brandt, K. "Should osteoarthritis be treated with nonsteroidal anti-inflammatory drugs?" *Rheumatic Diseases Clinics of North America* 19(3) (August 1992): 697–712.

Brandt, K. "Effects of nonsteroidal anti-inflammatory drugs on chondrocyte metabolism in vitro and in vivo." *American Journal of Medicine* 83(5a) (20 November 1987): 29–35.

Dingle, J. T. "Cartilage maintenance in osteoarthritis: Interactions of cytokines, NSAID and prostaglandins in articular cartilage damage and repair." *Journal of Rheumatology, Supplement* 28 (March 1991): 30–37.

Glorioso, S., Todesco, S., et al. "Impact on Hips and Knees." *International Journal of Clinical Pharmacology Research* 5(1) (1985): 39–49.

Hammand, M. F., Vilamitjana, J., Maloche, E., et al. "Effects of S-adenosylmethionine on human articular chondrocyte differentiation. An in vitro study." *American Journal of Medicine* 83(5A) (20 November 1987): 48–54.

Harrar, Sari, and O'Donnell, Sara Altshul. *The Woman's Book of Healing Herbs*. Emmaus, PA: Rodale Press, 1999.

Henrotin, T., Bassleer, C., and Franchimont, P. "In vitro effects of etrodolac and acetylsalicylic acid on human chondrocyte metabolism." *Agents & Actions* 36(3–4) (July 1992): 317–323.

Konig, B. "A long-term (two years) clinical trial with S-adenosylmethionine for the treatment of osteoarthritis. *Am J Med* 83(5A) (1987): 89–94.

Lewandowski, B., Bernacka, K., et al. "Piroxicam and posteroidal damage of articular cartilage." *Roczniki Akademii Medycznej W Bialymstoku* 40(2) (1995): 396–408.

Manicourt, D. H., Druetz-Van Egersen, A., Haazen, L., and Nagant De Deux-chaisnes, C. "Effects of tenoxicam and aspirin on the metabolism of proteoglycans and hyaluronan in normal and osteoarthritic human articular cartilage." *British Journal of Pharmacology* 113(4) (December 1994): 1113–1120.

Moskowitz, Roland, W. "Primary osteoarthritis: Epidemiology, clinical aspects, and general management." *American Journal of Medicine* 83(Suppl 5A) (20 November 1987).

Rovati, L. C. *International Journal of Tissue Reactions* 14(5) (1992): 243–251.

Shield, M. J. "Anti-inflammatory drugs and their effects on cartilage synthesis and renal function." *European Journal of Rheumatology & Inflammation* 13(1) (1993): 7–16.

6

Easing the Ache: SAM-e and Fibromyalgia

Imagine having a disease that has lots of symptoms, some of which are chronic and debilitating, but for which there are no tests to definitively detect them. Imagine that every muscle in your body aches, most or all of the time, yet no one can find a cause or reason for it. Imagine feeling profoundly tired or fatigued every day of your life, even after sleeping through the night.

If you are one of the estimated 3.7 million people with fibromyalgia, you don't need to imagine any of these symptoms, because you live with them every day. (This figure comes from the National Arthritis Data Workgroup; some experts say the figure is as high as 10 million.) Fibromyalgia is a hidden disease. On the outside, people with fibromyalgia often look healthy. On the inside, there is a physical war raging, causing pain, stiffness, and tenderness of the muscles and joints. There is also a mental battle going on, as depression, anxiety, and problems with concentration and short-term memory seem to take over their brain.

The trouble is, there are no lab tests, no X rays, no diagnostic "alarms" in the blood, urine, or other body fluids that say "Fibromyalgia happening here." In fact, even though fibromyalgia is classified as a rheumatic disorder, tests on the muscles of people with fibromyalgia, including muscle biopsy and electromyography (a test that measures the electrical activity of muscles), do not reveal any inflammation or abnormalities of the muscle. Whatever is causing the pain is something medical science does not have the technology or understanding to see.

That's when doubt can start to set in. You know you're exhausted and hurt all over, but everyone else, including your doctors, is telling you it's "in your head." "The tests are all negative; there's nothing wrong with you," they say. So you start to believe them . . . for a while. But you know you're not imagining things. You know there's something wrong. Often, that something is fibromyalgia.

Research into the role SAM-e can play in the treatment of fibromyalgia is relatively new, yet some of the findings thus far are promising. Others, while they don't demonstrate wildly positive results, also do not reveal any negative effects from using SAM-e. First, let's take a closer look at fibromyalgia, and then at how SAM-e is helping those with the disease.

What Is Fibromyalgia?

Since the early 1900s, there have been reports of people suffering with symptoms now recognized as fibromyalgia. Yet it was not until the 1980s that doctors and others in the medical arena began to look more closely at the symptoms as a whole and to treat them. A name was assigned to the condition—fibrositis—which means "inflammation (itis) of the fibrous tissue (fibro)." That name, however, did not accurately describe the primary symptoms of the disease.

Therefore, in 1990, a disease criteria committee at the American College of Rheumatology changed the name to fibromyalgia (*algia* is Greek for "pain"; *mys* is Greek for "muscle"; and *fibro* is Latin for "fibrous tissue"), and identified the major criteria for diagnosing the disease. For a diagnosis of fibromyalgia to be made, a patient must meet all three of the major criteria, which are: (1) a history of pain (three months or longer) in at least three anatomical sites in the body; (2) pain in at least eleven of eighteen sites designated as "tender points"—sites on the body that are painful when pressed; and (3) exclusion of other conditions that can cause similar symptoms. The tender points are located at the base of the skull, between and above the shoulder blades, below the elbows, in the lower back, behind the knees, and on the hips.

Several years after these criteria were established, thirty-five experts on fibromyalgia decided that these guidelines, which were created for research purposes, were not adequate for diagnosing the people they were seeing in their practices. So they determined that people did not have to have all eleven tender points to warrant a diagnosis of fibromyalgia if they had widespread pain and many of the common symptoms associated with the disease. These symptoms include generalized fatigue; numbing or tingling sensations in the hands, arms, legs, or feet; sleep disturbances; jaw pain (temporomandibular joint dysfunction); malaise and muscle pain after physical exertion; skin sensitivities; morning stiffness; irritable bowel; chronic headache (either tension or migraine); memory impairment (short-term memory loss) and/ or difficulty concentrating; menstrual cramping and PMS (premenstrual syndrome); and dizziness or impaired coordination. Most physicians are quick to add depression or the "blues" to this list, as well as a tendency for symptoms to vary greatly based on weather changes, activity, and stress. Some of the criteria are discussed in more detail below.

One problem that arises when trying to define fibromyalgia is that it is a systemic condition, meaning that it involves several bodily systems. In the case of fibromyalgia, those systems are the central nervous system, the endocrine system, the immune system, and, some researchers believe, also the limbic system (a subsystem of the central nervous system). This characteristic of fibromyalgia makes it difficult to study, treat, and manage.

Symptoms of Fibromyalgia

The symptoms of fibromyalgia usually vary greatly among people who have the disease, as well as from day to day for each person. The only thing you can count on, says one woman with fibromyalgia, is the pain. A detailed description of the pain, and other symptoms, is below.

Pain

The pain of fibromyalgia may begin insiduously, often starting in the neck and shoulders, and then gradually spreading throughout the body. The pain can be sharp or like a muscle cramp; dull and persistent or stabbing and transient. Sometimes the pain seems to move from place to place. Most people with fibromyalgia are in pain all the time to some degree. They note that the pain may change intensity or location according to the weather, time of day, level of activity, amount of stress in their lives, or the amount of sleep they get. SAM-e can be helpful in relieving the pain associated with fibromyalgia, as is discussed later in this chapter.

Fatigue and Sleep Problems

Fatigue is a major problem for about 90 percent of people with fibromyalgia. One source of fatigue is the abnormal sleep patterns that affect these individuals. Most people with fibromyalgia have chronic insomnia. Not only do they have trouble falling asleep, even though they are always tired, but they don't stay asleep. If they do sleep through the night, they still wake up exhausted. Scientists believe this occurs because people with fibromyalgia do not get sufficient "delta sleep." Delta sleep is the deep stage of sleep that occurs after people have been asleep for some time, usually about 90 minutes. The word "delta" refers to the brain waves (delta) that scientists have seen and mapped in people they have tested during sleep studies. Delta sleep is the most restorative stage of sleep. Without it, people wake up feeling as though they did not sleep at all.

Depression and Mental Fatigue

Although only about 25 percent of people with fibromyalgia are diagnosed as being clinically depressed, most say they feel depressed, blue, or anxious much of the time. Researchers have not determined

whether depression and anxiety are a result of fibromyalgia or a cause of it. Perhaps it is both. Mental
fatigue—problems with concentration, difficulty per-
forming simple mental tasks, and loss of short-term
memory—is also a symptom of fibromyalgia.

Other Symptoms of Fibromyalgia

Many other symptoms affect people with fibromyalgia. Although they are not part of the minor criteria
list, they have been reported by a significant number
of patients. These symptoms often come and go, and
may get worse when these individuals are under
stress, become ill, or overly exert themselves physically. These symptoms include morning stiffness, a
frequent need to urinate, alternating constipation
and diarrhea, an intolerance to cold, dizziness, and
abdominal pain.

Fibromyalgia and Chronic Fatigue Syndrome

Many individuals with fibromyalgia also have symptoms of chronic fatigue syndrome, a condition characterized by persistent or recurrent fatigue that cannot
be attributed to any definitive cause or illness. In fact,
approximately 70 percent of patients with fibromyalgia have all the diagnostic criteria for chronic fatigue
syndrome. The only difference in the diagnostic criteria between these two disorders is the requirement of
musculoskeletal pain in fibromyalgia and fatigue in
chronic fatigue syndrome. However, most people
with fibromyalgia *do* have fatigue. Some research indicates there may be a genetic link between these two

conditions, although more investigative work needs to be done.

Who Gets Fibromyalgia?

Fibromyalgia usually strikes people between the ages of twenty and fifty-five, although children can also get the disease. Ninety percent of cases of fibromyalgia are women, usually in their early to mid thirties. In fact, women are ten times more likely to get the disease than men. In 1985, Muhammed B. Yunus, M.D., of the University of Illinois College of Medicine, was the first to present a published report of the disease in adolescents. Calling it juvenile fibromyalgia syndrome, he noted that children reported symptoms similar to those in adults, but had fewer tender points. Up to 28 percent of adults with fibromyalgia say they first had symptoms during childhood.

Investigators have identified some risk factors for fibromyalgia. People with other rheumatic diseases such as rheumatoid arthritis (a chronic disease that involves pain and inflammation of the joints) and lupus (an inflammatory disease that affects the connective tissues) are at higher risk of having fibromyalgia. About 20 percent of people with rheumatoid arthritis, for example, also have fibromyalgia, but researchers do not understand the connection between these two conditions. Although fibromyalgia sometimes affects more than one member of the same family, no hereditary link has yet been identified.

Problems with Diagnosis

As mentioned in the opening of this chapter, one problem with diagnosing fibromyalgia is that there are no tests that identify the disease. Another problem is that many of the tender points are similar or the same as those associated with other inflammatory or muscle-related conditions, such as bursitis and tendonitis. This can cause physicians to misdiagnose the real problem and prescribe treatment or medications for the assumed one. Yet another difficulty with diagnosis is that many doctors are not familiar with fibromyalgia and so do not know how to evaluate for tender points. Arthritis and rheumatic disease specialists (called rheumatologists) are usually more familiar with this examination. That's because fibromyalgia is classified as a type of arthritis, a "soft-tissue rheumatism," and so is much more likely to be recognized and properly evaluated by specialists in rheumatism.

Causes of Fibromyalgia

Investigators have still not identified the cause or causes of fibromyalgia, but they have some theories. One points to low serotonin levels, which decreases people's tolerance to pain and causes sleep disturbances. John Russell, M.D., Ph.D., assistant professor of medicine and director of the University Clinical Research Center at the University of Texas Health Sciences Center in San Antonio, found that serotonin may be low or poorly metabolized in people who have fibromyalgia. Michael Murray, N.D., author of *Encyclopedia of Natural Medicine* and *Encyclopedia of Nutritional Supplements*, says that "although fibromyalgia has many facets, the central cause of the pain of fibromyalgia is a low level of serotonin." This concept supports a role for SAM-e, because it helps increase and maintain serotonin levels in the brain.

Scientists are also looking to see if the high levels of substance P (a molecule in the spinal cord that produces pain) in people with fibromyalgia is a cause of their persistent pain. In one study, investigators checked the level of substance P in the spinal fluid of people with fibromyalgia over a period of months, then checked it again after applying pressure to tender points. The fact that there was no increase in substance P levels when applying pressure indicates that the substance P level is always high and may be due to a defective internal process.

Yet another cause may be linked to excessive secretion of hormones. Leslie Crofford, M.D., assistant professor of internal medicine at the University of Michigan in Ann Arbor, found that people with fibromyalgia have higher pituitary hormone activity and less adrenal hormone activity than people without fibromyalgia. These hormones are associated with how we respond to stress. Dr. Crofford believes the abnormal levels of these hormones may be tied to the fatigue, sleep problems, and difficulty in concentrating that affect people with fibromyalgia.

Scientists may find that one or more of these theories are correct, or they may uncover new ones. In the meantime, some research studies have found that SAM-e helps relieve the pain and depression of fibromyalgia. Evidence of their findings is presented later in this chapter under "SAM-e and Fibromyalgia."

Pharmacological Treatment of Fibromyalgia

Treatment of fibromyalgia focuses on reducing pain, depression, and sleep disturbances. Because inflammation is not a symptom of fibromyalgia, drugs such as the nonsteroidal anti-inflammatories acetamino-phen (nonaspirin painkillers) and corticosteroids (e.g., prednisone and cortisone) are generally not ef-fective. The little relief they may provide in the way of pain reduction is more than offset by the side ef-fects of each drug type. Because fibromyalgia is a chronic disease, use of narcotics, which are addictive, is not a good choice. This leaves fibromyalgia sufferers with little choice except antidepressants and tranquil-izers (and of course, SAM-e, which is discussed in the next section). These are discussed briefly below.

Antidepressants

When given in lower doses than used to treat depres-sion, tricyclic antidepressants such as amitriptyline, doxepine, and nortriptyline, can be effective in treat-ing fibromyalgia. (See Chapter 3 for details on tricy-clic antidepressants.) Low doses help relax muscles and promote better sleep for some people with fi-bromyalgia. The possible benefits must be weighed against the side effects, which can include daytime drowsiness, dry mouth, weight gain, and constipa-tion. There are mixed findings on the effectiveness of another type of antidepressant, the selective seroto-nin reuptake inhibitors, or SSRIs, on fibromyalgia. As their name implies, these drugs, which include fluox-etine (Prozac) and sertraline (Zoloft), help the body retain serotonin, but some studies did not find them useful. In addition, SSRIs can cause insomnia—a side effect fibromyalgia sufferers do not need.

Antianxiety Drugs and Sedatives

Antianxiety drugs and sedatives are sometimes pre-scribed to help produce a calming effect, reduce muscle pain, and induce sleep. Diazepam (Valium) can help relax the muscles, yet it is associated with drowsiness and a "hangover" feeling. Triazolam (Halcion) and temazepam (Restoril) are sometimes used to treat the insomnia, yet both can also cause dizziness and confusion.

SAM-e and Fibromyalgia

SAM-e is an exciting treatment for fibromyalgia because, in an arena where there are so few drugs that can offer any relief, SAM-e may provide benefits that equal and sometimes surpass those of conventional drugs, and without the side effects. And when you are treating a chronic condition, you need treatments that are safe. So far, SAM-e fits that requirement. Below is a review of a few studies that demonstrate the benefits of SAM-e in the treatment of fibromyalgia. Although they are not included, there are also several studies in which SAM-e did not provide significant relief of symptoms. However, because it has been effective in many cases and has thus far proven to be safe, many researchers and physicians believe SAM-e is worth trying in patients with fibromyalgia.

One of the first studies to evaluate SAM-e in the treatment of fibromyalgia was done in Italy in 1987. Seventeen people with primary fibromyalgia participated in what is called a "cross-over study," in which patients first receive one treatment and then are switched or "crossed over" to another treatment. In this study, one group received 200 mg of oral SAM-e daily and the other group received placebo. At the end of twenty-one days, both groups stopped taking any treatment for fourteen days. Then the two groups switched treatments. The researchers found that patients experienced a significant improvement in mood and a decrease in the number of tender points and other painful sites when they were taking SAM-e but not when they were taking placebo. The patients also scored better on depression tests, including the HAM-D, when taking SAM-e but not when taking placebo. These findings lead the doctors to state that "SAM-e . . . seems to be effective and safe therapy in the management of primary fibromyalgia."

A subsequent study in 1991 was performed in Denmark. There, Dr. S. Jacobsen and his research team conducted a double-blind, placebo-controlled study of forty-four patients with primary fibromyalgia. Half of the patients were given 800 mg of oral SAM-e daily and the other group was given a placebo. The study lasted six weeks, at the end of which time the patients who had received SAM-e experienced an improvement in fatigue, morning stiffness, pain, and mood when compared with patients who had received placebo. Overall, the researchers reported that SAM-e "has some beneficial effects on primary fibromyalgia and could be an important option in the treatment hereof."

A 1994 study in Italy also lasted six weeks. In this uncontrolled trial, forty-seven patients with primary fibromyalgia were given SAM-e. The investigators reported that SAM-e "proved to be effective in relieving pain and impaired mood" and also found that it improved sleep quality.

Another double-blind, placebo-controlled, crossover study was done by a research team in Denmark in 1997. Thirty-four patients were given 600 mg of intravenous SAM-e or placebo daily for ten days.

Treatment with SAM-e resulted in a less than significant improvement in sleep, fatigue, morning stiffness, pain at rest, and pain when moving than treatment with placebo. Both groups had a similar response to sensitivity to pain at tender points. Overall, the researchers did not consider this to be a positive study. One reason for the unsatisfactory results could be the very short length of the study. All other reported studies have lasted at least twenty-one days. The fact that there was some slight improvement in some of the symptoms indicates that had the study been continued for a longer period of time, the results may have been significant.

Not every controlled study of SAM-e has compared its effectiveness with placebo or a drug. Dr. DiBenedetto and his research team evaluated how SAM-e performed against a therapy method called TENS, or transcutaneous electrical nerve stimulation. A TENS therapy unit is a small device about the size of a pager which has thin wires that carry mild electrical pulses directly to painful sites. In Dr. DiBenedetto's six-week study, fifteen patients with primary fibromyalgia received either a 200-mg injection of SAM-e in the morning, plus one 200-mg tablet at noon and in the evening; or they were given a TENS unit to wear. All the patients underwent testing on three depression scales. Patients who received SAM-e showed a significant reduction in depression scores on all three scales, while those with a TENS unit scored much lower. When five of the TENS patients switched over and took SAM-e, their depression scores decreased as well. Although the patients did not directly experience a reduction in other symptoms of fibromyalgia, the significant improvement in depression scores was considered to be of major benefit to the patients.

Other Nondrug Treatments for Fibromyalgia

Several alternative medications, including 5-HTP, St.-John's-wort, and magnesium have proven effective for some people with fibromyalgia. (See Chapter 9 for information on how to take these remedies with SAM-e.) Other natural treatments that can be effective in relieving symptoms of fibromyalgia include acupressure, acupuncture, relaxation techniques, therapeutic massage, physical therapy, and gentle exercise, including tai chi and yoga. These therapies are not within the scope of this book; however, you are encouraged to contact one of the organizations concerned with fibromyalgia and ask about whether these treatments may help you (see the Appendix).

A combination of St.-John's-wort, 5-HTP, and magnesium has been recommended by Michael Murray, N.D., and others, for the treatment of fibromyalgia. They base this natural treatment recommendation on several studies that show St.-John's-wort and 5-HTP (L-5-hydroxytryptophan) to be effective individually for treatment of depression and fibromyalgia, and that 5-HTP produces significantly better results when combined with another antidepressant. However, each of these remedies provides significant results in its own right. (See Chapter 9 for the recommended dosages of these three supplements.)

5-HTP. The natural substance 5-hydroxytryptophan, or 5-HTP, is valued as an antidepressant, and was discussed in detail in Chapter 4. In addition to its antidepressant benefits, however, 5-HTP has the ability to improve the quality of sleep, morning stiffness, fatigue, anxiety, and the number of painful areas in people with fibromyalgia. According to a recent study by Federigo Sicuteri of the University of Florence, "In our experience, as well as in that of other pain specialists, 5-HTP can largely improve the painful picture of primary fibromyalgia."

St.-John's-wort. This herb also is valued as an antidepressant (see Chapter 4 for details). The reason it is being combined with 5-HTP here is that a study showed how 5-HTP, when taken with a conventional antidepressant, produced significantly better results than if either substance had been taken alone. In the combination suggested here, St.-John's-wort takes the place of the conventional drug.

Magnesium. This mineral is found in low levels in people with fibromyalgia. Magnesium is a very active substance and is critical for many cell functions, including energy production and the processes involved in the manufacturing of ATP. Therefore if magnesium levels are low, energy levels are low.

What is SAM-e's future role to be in the treatment of fibromyalgia? So far the studies have provided mixed results, although none of them have been negative. Clearly more, larger, and longer studies are needed before anyone can pass judgment on SAM-e's effectiveness in the treatment of fibromyalgia.

NOTES

Arthritis Foundation. *Your Personal Guide to Living Well with Fibromyalgia*. Atlanta: Longstreet Press, 1997.

Byerley, W. F., et al. "5-hydroxytryptophan: A review of its antidepressant efficacy and adverse effects." *J Clin Psychopharmacol* 7 (1987): 127–137.

Caruso, I., et al. "Double-blind study of 5-hydroxytryptophan versus placebo in the treatment of primary fibromyalgia syndrome." *J Int Med Res* 18 (1990): 201–209.

Cote, K. A., et al. "Sleep, daytime symptoms, and cognitive performance in patients with fibromyalgia." *J Rheumatol* 24(10) (October 1997): 2014–2023.

De Smet, P. A. G., and Nolen, W. "St. John's wort as an antidepressant." *BMJ* 313 (1996): 241–242.

DiBenedetto, P., Iona, L. G., and Zidarich, V. "Clinical evaluation of S-adenosyl-O-methionine versus transcutaneous electrical nerve stimulation in primary fibromyalgia." *Curr Ther Res* 53 (1993): 222–229.

Godfrey, R. G. "Guide to the understanding and use of tricyclic antidepressants in the overall management of fibromyalgia and other chronic pain syndromes." *Archive of Internal Medicine,* 27 May 1996.

Grassetto, M., and Varotto, A. "Primary fibromyalgia is responsive to S-adenosyl-l-methionine." *Curr Ther Res* 55 (1995): 797–806.

Harding, S. M. "Sleep in fibromyalgia patients: Subjective and objective findings." *Am J Med Sci* 316(6) (June 1998): 367–376.

Harrer, G., and Schulz, V. "Clinical investigation of the antidepressant effectiveness of *Hypericum.*" *J Geriatr Psychitry Neurol* 7 (Suppl 1) (1994): S6–8.

Jacobsen, S., Danneskiold-Samsoe, B., and Andersen, R. B. "Oral S-adenosylmetionine in primary fibromyalgia. Double-blind clinical evaluation." *Scand J Rheumatol* 20(4) (1991): 294–302.

Morazzoni, P., and Bombardelli, E. "*Hypericum perforatum,*" *Fitoterapia* 66 (1995): 43-68.

National Arthritis Data Workgroup, unpublished data, 1997.

Nicolodi, M., and Sicuteri, F. "Eosinophilia myalgia syndrome [food and drink]: The role of contaminants, the role of serotonergic set up." *Exp Biol Med* 398 (1996): 351–357.

Older, S. A., et al. "The effects of delta wave sleep interruption on pain thresholds and fibromyalgia-like symptoms in healthy subjects; Correlations with insulin-like growth factor I." *J Rheumatol* 25(6) (June 1998): 1180–1186.

Poldinger, W., Calanchini, B., and Schwarz, W. "A functional-dimensional approach to depression: Serotonin deficiency as a target syndrome in a comparison of 5-hydroxytryptophan and fluxvoxamine." *Psychopathology* 24 (1991): 53–81.

Puttini, P.S., and Caruso, I. "Primary fibromyalgia syndrome and 5-hydroxy-L-tryptophan: A 90-day open study." *J Int Med Res* 20 (1992): 182–189.

Roizenblatt, S., et al. "Juvenile fibromyalgia. Clinical and polysomnographic aspects." *Journal of Rheumatology* 24(3) (March 1997): 579–585.

Russell, I. J., Michalek, J. E., Vipraio, G. A., et al. "Platelet 3H-Imipramine uptake receptor density and serum serotonin levels in patients with fibromyalgia/fibrositis syndrome." *Journal of Rheumatology* 19 (1992): 104–109.

Russell, I. J., Vaeroy, H., Javors, M., and Nyberg, F. "Cerebrospinal fluid biogenic amine metabolites in fibromyalgia/fibrositis syndrome and rheumatoid arthritis." *Arthritis and Rheumatism* 35 (1992): 550–556.

Shaver, J. L., et al. "Sleep, psychological distress, and stress arousal in women with fibromyalgia." *Res Nurs Health* 20(3) (June 1997): 247–257.

Touchon, J. "Use of antidepressants in sleep disorders: Practical considerations." *Encephale* no. 7 (21December 1995): 41–47.

Tavoni, A., Vitali, C., Bombardieri, S., and Pasero, G. "Evaluation of S-adenosylmethionine in primary fibromyalgia. A double-blind crossover study." *American Journal of Medicine* 83(5A) (November 20, 1987): 107–110.

Van Hiele, J. J. "L-5-hydroxytryptophan in depression: The first substitution therapy in psychiatry?" *Neuropsychobiology* 6 (1980): 230–240.

Van Praag, H. M. "Management of depression with serotonin precursors." *Biol Psychiatry* 16 (1981): 291–310.

Volkmann, H., Norregaard, J., Jacobsen, S., et al. "Double-blind, placebo-controlled cross-over study of intravenous S-adenosyl-L-methionine in patients with fibromyalgia." *Scand J Rheumatol* 26(3) (1997): 206–211.

7

The Great Detoxifiers: SAM-e and Liver Disease

Bile. Cholesterol. Toxins. Fat. These words do not stir up pretty images. Now let's add another word: liver. Another word that conjures up unattractive thoughts. But if it were not for this rather unappealing organ, your body would have a big problem with all of the not-so-pleasant substances just mentioned.

A healthy liver is essential for life. SAM-e has a vital role in keeping the liver healthy, because the liver depends on an adequate supply of SAM-e to perform its many functions. One fact about SAM-e serves as a big clue as to its importance to the liver: of the approximately eight grams of SAM-e that the body produces each day, more than half of that amount is made in the liver. When supplies of SAM-e are low, trouble follows. When researchers discovered that SAM-e supplements can be beneficial to the liver, it stirred up some excitement in the research world. "Perhaps nowhere are SAMe's benefits so dramatic as in the liver," says Sol Grazi, M.D., author of *The European Arthritis and Depression Breakthrough! SAMe*.

This chapter reveals those benefits by telling you about the research and studies behind the use of SAM-e in liver disease. If you or someone you care about has cirrhosis, hepatitis, or another liver disease, this chapter will be of interest to you. Even if you don't have a personal interest in liver disease, the story of how the liver works and how SAM-e enhances its functioning is important to everyone, because without a liver, you could not exist.

The Liver Story

The liver is one of the most complex and fascinating organs in the body. Weighing in at about three pounds in adults and taking up the space of a football, it is located behind the lower ribs on the right side of the abdomen. From that location, it performs an impressive number of functions.

- It converts food into chemicals the body uses for growth and life functions.
- It manufactures and disperses vital substances the body needs, including proteins, urea, and cholesterol.
- It processes drugs absorbed from the digestive tract and chemically modifies them into forms the body can use.
- It detoxifies and eliminates substances from the bloodstream that can be poisonous to the body and discharges the waste products into the bile.
- It stores iron, sugars, and some vitamins until they are needed by the body.

As you look at the functions listed here, it's easy to imagine the liver as a central processing plant. All of the blood that leaves the stomach and intestines must circulate through the liver before it is sent to other destinations in the body. Everything you ingest—food, beverages, alcohol, drugs—plus the air you breathe and substances that are absorbed through your skin; everything, in some form, is processed through the liver. Whether it is converting, manufacturing, dispersing, detoxifying, or storing substances, the liver is always at work.

Each of the liver's functions is critical to life, and so an understanding of them will help you appreciate the role SAM-e has in preserving, maintaining, and promoting a healthy liver.

The Converting Liver
The liver is responsible for storing fats (as fatty acids) and carbohydrates (as glycogen, which is a complex carbohydrate composed of glucose molecules) from the food you eat. The liver can easily convert glycogen back to glucose and release it into the blood when it is needed for energy. If you consume more carbohydrates than the body needs or can be stored as glycogen, the liver converts the excess into fat. This fat is then stored and used in several processes, including manufacturing of cholesterol.

The Manufacturing Liver
The liver manufacturers many substances necessary for life. One is cholesterol, which despite all the negative publicity it gets, is not all bad. Cholesterol is manufactured from the stored fatty acids in the liver. It is involved in metabolism and is a precursor of sev-

eral steroid hormones, including the sex hormones. Cholesterol is an ingredient in bile, which is one place it has a positive function, but it also has several very negative roles. One is its tendency to accumulate in the blood vessels, where it restricts the flow of blood. Another place you find cholesterol is in gallstones. Cholesterol is the primary ingredient of most gallstones. Gallstones form when the bile contains more cholesterol than it can manage. Approximately 20 percent of women and 8 percent of men older than age forty have gallstones, although some people don't know it because they have no symptoms. Apparently enough people do have symptoms, however: every year more than 300,000 people have their gallbladder removed because of gallstones.

The liver also manufactures bile. This thick, straw-colored substance is composed of pigments, cholesterol, and various organic and nonorganic materials. After the bile is produced, it is stored in the gallbladder which, after you eat, discharges the bile into the intestine. About one quart of bile enters the small intestine each day. There it assists in the absorption of oils, fats, and fat-soluble vitamins, such as vitamins A, D, and E. Bile also helps keep the small intestine healthy and free from microorganisms, and is instrumental in ensuring that water is incorporated into the stool. Insufficient bile can result in hard stools and constipation.

Other substances manufactured by the liver include various proteins, including albumin, fibrinogen, and transferrin, and urea, a nitrogen compound found in urine.

The Processing and Modifying Liver

When drugs are made available to the market, they have already been tested for how they interact with the liver and how long it takes for the liver to process and modify the drug. Thus, when you take medication, the liver is standing by, ready to process and modify it. If you have liver disease, the liver's ability to filter and modify drugs is hindered, which allows the toxins in the drugs to remain in the bloodstream longer than normal. The toxins can then accumulate and affect the brain, causing mental confusion or even changing an individual's personality. Because the drugs were designed to stay in the body only for a specified amount of time, the fact that they are lingering in the bloodstream longer than expected means people can have adverse reactions to the drugs as well as experience severe side effects. Some drugs can be especially damaging to the liver, including the over-the-counter drug acetaminophen.

The Detoxifying Liver

Much of the work the liver does involves detoxifying potentially dangerous substances and then eliminating them from the body or finding a way for them to be used. The body is constantly exposed to toxins, twenty-four hours a day, both from the outside environment and from those manufactured within the body as the byproducts of metabolism and other bodily functions. If the liver is unable to break down the toxins and wastes that pass through it, they will get into the bloodstream and slowly poison the body to death.

Toxins from the outside environment enter the

body through food, water, the air, and the contents of the bowel. It is impossible to completely eliminate your exposure to toxins. Pesticides and herbicides, for example, are used on most food crops in the United States and other countries. Even if you wash fruits and vegetables well or peel them, residue still gets into the food. Organic foods are an improvement, but even they have naturally occurring toxic components. Cows, chickens, pigs, and other food animals eat feed that has been treated with hormones, pesticides, steroids, and other poisons. These toxins linger in their muscles and show up on your plate. Gasoline fumes, cigarette smoke, chemical pollutants, and other airborne poisons are all around you, riding the molecules as you breathe them in.

Even if you locked yourself in a sterilized bubble, your body would still be producing a long list of toxins that are the byproducts of various metabolic and other chemical processes that occur in the body all the time. When the liver is healthy and internal toxin levels are normal, the liver can neutralize these internal poisons. If any metabolic processes are disrupted, usually because of nutritional deficiencies, then levels of some toxins may accumulate in the body and cause disease or other medical conditions.

But the liver has a highly specialized detoxification process to take care of these poisons. If you were to take a slice of your liver and examine it under a microscope, you would see rows of liver cells separated by spaces. These spaces, called sinusoids, act as a sieve to filter the blood and remove the toxins and other waste products. The sinusoids are helped by special cells called Kupffer cells, which destroy all the material sent to them by the sinusoids. Thus the liver is a highly evolved waste disposal organ, with sophisticated chemical and enzyme processes, or pathways, that metabolize every particle that passes through the bloodstream.

The Liver's Detoxification Methods

The liver has four mechanisms it uses to detoxify the body. They include (1) blood filtration; (2) bile production; (3) phase one detoxification reactions; and (4) phase two detoxification reactions.

Blood Filtration. Every minute of every day, nearly two quarts of blood pass through the liver for filtration and detoxification. When the liver is functioning optimally, it can remove 99 percent of the toxins from the blood before it goes back into general circulation. When the liver is damaged or overburdened, toxins get back into the system without being filtered. This can be dangerous, especially when the blood comes from the intestinal tract. Blood from the intestines is typically overflowing with bacteria and various toxins that are the byproducts of different bodily processes. The potential for infection and disease is greatly increased when this toxic blood is circulated through the body.

Bile Production. The approximately one quart of bile that the liver manufactures each day carries many of the toxic substances that are to be eliminated from the body. These poisons are sent to the intestines, where they are absorbed by fiber and excreted in the stools. If a person's diet is low in fiber, some of the toxins can be reabsorbed by the body, or bacteria in the intestines may interact with the toxins and change them into substances that are even more damaging.

124

Phase One. During phase one, the liver takes the toxins and, through a complicated process, reduces them to less damaging substances. Some of the substances that are neutralized during phase one include bacteria and bacterial products, cholesterol, extra calcium, many prescription and over-the-counter drugs, caffeine, hormones, and some insecticides. The group of more than one hundred enzymes largely responsible for phase one detoxification is called the *cytochrome P450 system.* Each enzyme has its own types of toxins it can detoxify best, although there is much overlap among the enzymes, which work like a system of checks and balances. It is worth noting here that the cytochrome P450 enzyme system is also found in other parts of the body, especially in the brain. Low levels of antioxidants and nutrients in the brain can cause neuron damage, which is characteristic of Alzheimer's disease and Parkinson's disease, two other conditions that may be helped by SAM-e (see Chapter 8 for information on Alzheimer's disease and Parkinson's disease).

Many substances can inhibit the work of the cytochrome P450 process and thus increase the chances that the toxins will stay in your body longer and cause more damage. These substances are noted in the sidebar. At the same time, there are substances that help the phase one detoxification process. These are also noted in the sidebar.

The phase one detoxifying process produces byproducts, called oxygen free radicals, which can cause substantial damage to the liver unless there are antioxidants around to eliminate them. The most important antioxidant for neutralizing the free radicals produced during phase one is glutathione. Other es-

Substances and Conditions That May Inhibit and Activate Phase One Detoxification

Inhibitors

DRUGS: antihistamines; benzodiazapines (e.g., Librium, Valium); cimetidine and other drugs used to block stomach acid; ketoconazole; sulfaphenazole

FOODS: Capsaicin (in red chili peppers); curcumin (in the spice turmeric); eugenol (in clove oil); naringenin (in grapefruit and grapefruit juice)

AGING

TOXINS from bacteria in the intestinal tract

Activators

FOODS: Broccoli, cabbage, and brussels sprouts; oranges and tangerines

NUTRIENTS: niacin, vitamin B_1, vitamin C

HERBS: caraway and dill seed

sential antioxidants include vitamin C, selenium, and folic acid. Glutathione has been mentioned several times before, because SAM-e is a precursor for this essential antioxidant. It is needed in both phase one and phase two detoxification. If the level of toxins becomes so great that too many free radicals are produced to be handled by glutathione during phase one, the supply of glutathione can be used up, which means any phase two processes that depend on glutathione will not be possible.

Phase Two. During phase two, the toxins that were made less dangerous during phase one are now linked with a glycine or sulphate molecule, which makes them water soluble. This makes them easy to excrete from the body through the urine and bile. Phase two is also the time when any toxins not neutralized during phase one are now detoxified. Some of these include acetaminophen, nicotine, some insecticides, aspirin, benzoate (a common food preservative), estrogen, and morphine. The nutrients that aid phase two detoxification include glutathione, vitamin E, carotene, sulfur-containing amino acids (such as taurine, cysteine, methionine), glutamine, choline, and inositol.

If the detoxification processes during phase one or two become overburdened with toxins or the sinusoids or pathways are blocked, toxins will accumulate in the body. Many toxins are fat soluble, so they find their way into the fatty cell membranes and take up residence. Toxins such as pesticides, synthetic sugars, food additives, and petrochemicals are very toxic to the nervous system and in many cases are cancer causing. Buildup of these toxins in the brain, prostate, breast, and other body sites has been implicated in the increase in cancer. Accumulation of harmful microorganisms in the blood can overload the immune system and can result in recurrent infections, swollen glands, chronic fatigue, inflammatory disorders, allergies, and autoimmune diseases (diseases in which the body's immune system attacks itself; rheumatoid arthritis is one example).

The Comeback Organ

Nature, perhaps as a way of compensation for the continuous barrage of deadly toxins, has given the liver a unique ability: It can regenerate itself. No other organ in the human body can restore itself. As long as the structural integrity of the liver is not damaged, up to 80 percent of it can be removed and it will regenerate itself within a few months. If, however, the structure is damaged by toxins or disease, the rejuvenation process cannot occur.

Liver Disease: Cirrhosis

When chronic diseases damage the liver and cause it to become permanently scarred, the condition is known as cirrhosis. Scar tissue in the liver affects the organ's structure and blocks the flow of blood. The more scar tissue there is, the more the liver cannot function properly. Cirrhosis disrupts and can eventually destroy the ability of the liver to detoxify the blood; to produce bile, urea, cholesterol, proteins, and other vital substances; and to perform its many other functions properly.

Even though the liver has the ability to rejuvenate itself, sometimes the damage is too extensive. Approximately 25,000 people die of cirrhosis each year. It is now the seventh leading cause of death by disease in the United States.

According to the American Liver Foundation, liver diseases are on the rise. Part of the reason for the increase may be that we are exposed to more toxins more frequently than were past generations, especially in the food and water supply. As the liver has more and more toxins to process, it becomes overworked and breaks down.

Despite continuing research into liver disorders, there is much scientists and physicians still do not understand about the more than one hundred liver diseases. In addition, few effective treatments, except for liver transplants, have been found for most of the life-threatening liver disorders. These reasons make the potential for SAM-e to be an effective treatment for cirrhosis very exciting. Some of the studies that are leading researchers in that direction are discussed later in this chapter.

Causes of Cirrhosis

More than forty different conditions can cause cirrhosis. Many of them are rare, leaving just a few to be responsible for the majority of cases. In the United States, the most common cause of cirrhosis is alcoholism. Seventy-five percent of all cases of cirrhosis in the United States are attributed to overconsumption of alcohol, with twice as many men as women having the disease. Because it is such a significant cause of cirrhosis, it deserves explanation here.

Regular alcohol consumption, whether it is beer, wine, or liquor, over many years can cause cirrhosis. There is no "magic number" of drinks or years drinking them that guarantees you will get cirrhosis, because everyone's body chemistry and lifestyle is different. But if you or someone you know is drinking three or more cans of beer, four or more glasses of wine, or four or more shots of liquor a day for ten or more years, chances are good the liver on the receiving end of the alcohol has or will soon have cirrhosis. Even substantially less alcohol consumption causes some liver damage.

Alcohol does several things to the liver in particular and to the body in general as the liver finds it harder and harder to function properly. When alcohol enters the liver for processing, the liver converts the ethanol in the drink into acetaldehyde and then into vinegar. Acetaldehyde is toxic and damaging to the liver, but in small amounts the liver can handle it well. When the liver is presented with an excessive amount of acetaldehyde over a long period of time, the damage accumulates. Acetaldehyde inhibits the liver's ability to convert its stores of glycogen and fatty acids into energy for the body to use. Instead, the fatty acids and glycogen remain in the liver, exert pressure on the cells, and damage the liver. This creates a condition known as "fatty liver," the beginning stages of cirrhosis. Acetaldehyde also disrupts the liver's ability to convert fats into energy.

Unless the individual stops drinking at this stage, the damage will continue. As more and more liver cells are broken down, scar tissue forms in the liver and disrupts the flow of blood and other fluids through the blood vessels and bile ducts. Many symptoms and complications usually begin to appear at this point (see below, "Symptoms of Cirrhosis"), with consequences that ultimately can affect nearly every part of the body. If the damage to the liver becomes too extensive, the liver itself becomes a toxin to the body. At this stage, called alcoholic hepatitis, the individual may need a liver transplant or, more likely, will die.

Chronic viral hepatitis (types B, C, and D) is another common cause of cirrhosis. The B, C, and D designations refer to the type of virus that causes the disease. Chronic viral hepatitis is defined as liver inflammation caused by a virus. The disease may be mild, or so severe that it is fatal. Initial symptoms include low-grade fever, anorexia, nausea, vomiting, fatigue, muscle and joint pain, headache, cough, and light sensitivity. These symptoms typically last up to fourteen days in acute cases of viral hepatitis, but in some cases they continue and become chronic. Jaundice is usually the next symptom to appear, and it is often preceded by dark urine and clay-colored stools. As the hepatitis virus continues to attack the liver, the symptoms increase in severity, and the liver continues to be damaged, leading to cirrhosis.

Some causes of cirrhosis are inherited. The following conditions fall into that category:

- **alpha-1 antitrypsin deficiency**: a condition characterized by the lack of an enzyme that is associated with progressive cirrhosis in children and early onset emphysema in adults.
- **cystic fibrosis**: a disease of the exocrine glands (glands whose secretions reach the skin's surface, such as a sweat gland). It affects the pancreas, respiratory system, and the apocrine glands.
- **galactosemia**: a disease in which the absence of a specific enzyme prevents the body from converting galactose to glucose. Infants with galactosemia fail to thrive unless galactose and lactose are removed from the diet. Normally, the liver needs galactose to convert into glycogen.
- **glycogen-storage diseases**: any one of several conditions (designated by "type I, II, III, etc.) in which abnormal amounts of glycogen are stored in the tissues, and especially in the liver.
- **hemochromatosis**: a disease in which too much iron is absorbed and then stored in the liver (which becomes enlarged), pancreas, skin, intestinal lining, heart, and endocrine glands.
- **Wilson's disease**: a disorder in which the body stores too much copper in the liver, brain, kid-

neys, and the corneas of the eyes. In addition to cirrhosis, the disease is characterized by degeneration of the brain, tremors, and muscle rigidity.

Another cause of cirrhosis is a blocked bile duct. The bile ducts carry the bile from the liver to the intestines. Cirrhosis in infants is usually caused by a disease called biliary atresia, in which the bile ducts are blocked, injured, or absent. In adults, the bile ducts can become blocked, inflamed, or scarred due to primary biliary cirrhosis or after gallbladder surgery in which the bile ducts are tied off or damaged. Less common causes of cirrhosis include prolonged exposure to environmental toxins and chemicals, a reaction to prescription or over-the-counter drugs (e.g., acetaminophen can cause liver damage when taken in large doses for a prolonged period of time), and repeated episodes of heart failure with liver congestion.

Symptoms of Cirrhosis

People with cirrhosis may have the disease for some time before they notice anything unusual. The liver is a hearty organ and can take much abuse before it causes outward signs of trouble. As the liver loses functioning cells and liver scarring becomes extensive, fatigue, weakness, and exhaustion are usually the first symptoms. These are typically followed or accompanied by a loss of appetite, nausea, and weight loss. As the liver loses its ability to produce proteins, other symptoms appear. A decrease in the production of the proteins required for blood clotting results in a tendency to bleed or bruise easily. Low levels of the protein albumin causes water to accumulate in the legs or abdomen. As the disease progresses, the skin and the whites of the eyes become yellow, a condition known as jaundice. Jaundice is the result of an accumulation of bile pigments that are transported to the intestines. Intense itching occurs in people who get a buildup of bile in their skin, and gallstones can form when an inadequate supply of bile reaches the gallbladder.

Because the liver's ability to detoxify the blood is hampered by the loss of liver cells and scarring to the organ itself, toxins begin to build up in the blood and the brain. Early signs of toxin accumulation in the brain are forgetfulness, difficulty concentrating, changes in sleeping habits, unresponsiveness, and neglect of personal appearance.

A person with cirrhosis who is taking any type of medication will notice that the drugs' effects will last longer. That's because the liver is no longer able to remove the drugs from the blood as quickly or efficiently as it did when the liver was healthy. Thus the drugs build up in the body and can make people very sensitive to both the drugs and any side effects they may cause.

Another common and serious problem for people with cirrhosis is a condition called *portal hypertension.* Because the blood flow in the liver is impaired in cirrhosis, the blood that flows into the liver from the intestines and spleen exerts too much pressure on the blood vessels that enter the liver. This causes blood to back up and enlarge the spleen. Blood from the intestines flows around the liver through new blood vessels. These new vessels may form in the stomach and can become enlarged (they are called varices) because of the increased pressure. Unfortunately, their walls are usually very thin, and the combination of thin walls and high pressure can cause them to rupture and bleed into the stomach or esophagus, which is a life-threatening situation.

Diagnosing Cirrhosis

Physicians usually diagnose cirrhosis using a combination of the patient's symptoms, a physical examination to detect if the liver is enlarged, and laboratory tests if cirrhosis is suspected from the symptoms and examination. Blood tests are usually taken; a computerized axial tomography (CAT) scan, radioisotope liver/spleen scan, or ultrasound also may be performed. Sometimes a physician will do a liver biopsy through the skin to get a tissue sample, or view the liver through a laparoscope, a minute viewing instrument that is inserted through a tiny incision in the abdomen.

Treating Cirrhosis

The purpose of any treatment of cirrhosis is to stop or slow its progress, reduce and relieve symptoms and complications, and minimize damage to the liver. Cirrhosis is associated with so many symptoms and complications that treatment can be quite a task. If the cause of the cirrhosis is alcoholism, the person must stop drinking for treatment to be effective. Because cirrhosis causes nutritional deficiencies, supplementation with a multivitamin-mineral and specific nutrients such as vitamin K, vitamin B_1, and folic acid is recommended. Protein intake should be restricted to help avoid toxic buildup in the digestive tract, and dietary salt should be reduced to deal with any fluid retention in the legs, ankles, and abdomen. Diuretics can also help remove excess fluid.

If the individual also has hepatitis, steroids or antiviral drugs are given to reduce further damage to the liver. The only FDA-approved drug for chronic hepatitis B or C is called interferon alpha-2b. This is an injectable drug and has many side effects, including headache, flu-like symptoms, and nausea.

When portal hypertension is a problem, antihypertensive medications, such as a beta-blocker, are often prescribed. If bleeding from the varices in the stomach occurs, a sclerosing agent (a substance that can strengthen the blood vessels) can be injected through a flexible tube that is inserted through the mouth and esophagus. If the damage becomes life-threatening, surgery is the only remaining alternative. Surgeons can insert a device called a shunt that can relieve the pressure in the portal vein. The only other choice is a liver transplant.

SAM-e and Liver Disease

Common sense tells us that if the liver has the highest concentration of SAM-e in the entire body, then SAM-e must have many roles to play in the liver. Researchers have discovered many of the functions SAM-e performs in the liver. One of them is to regulate the fat and fluid content of the liver cells; another is to improve the flow of bile. Yet another is its lipotropic effect, which means it prevents fats from building up in the liver, the condition known as "fatty liver" which leads to cirrhosis.

SAM-e also transfers sulfur, a mineral that is used to make glutathione, taurine, and cysteine, all of which are involved in metabolism. Both taurine and glutathione are needed to make bile salts.

People with liver disease have low levels of SAM-e, and the reason appears to be that cirrhosis damages the enzyme (called S-adenomethionine synthetase) that allows methionine to combine with ATP to make SAM-e. The inability to make SAM-e causes several chains of events. One, there is an excess of methionine, which increases the toxin load for the liver. Two, without sufficient SAM-e, the liver cannot manufacture glutathione, which is the main antioxidant the liver uses during its detoxification process. Three, without the taurine and glutathione that SAM-e produces, the liver cannot make bile salts, which damages its bile production, which reduces the body's ability to metabolize fats. Four, low levels of SAM-e in the liver hamper the organ's ability to maintain cell growth and cell repair.

Thus it appears that giving SAM-e to anyone with liver disease should stimulate some sort of response in the liver. Although a SAM-e supplement would not necessarily boost the meager levels of SAM-e that are already in the liver, it would allow the normal processes to continue—the production of glutathione, proteins, bile, taurine, and so on—as if SAM-e had been manufactured right there in the liver. Based on the studies of SAM-e in cirrhosis and other liver disorders, this assumption appears to be true.

SAM-e and Liver Disease: The Studies

The Early Studies

Because SAM-e has been in widespread use in Europe since the 1970s, it is only natural that early research took place there. (See Chapter 4 or the Glossary for an explanation of terms used to describe research studies.) In 1975, several studies were published on the use of SAM-e in cirrhosis and chronic hepatitis. In one double-blind study, conducted by Dr. Labo and his associates, fifty-three patients with cirrhosis were divided into two treatment groups. Twenty-eight patients received 150 mg of intravenous (IV) SAM-e daily plus 2,000 U of vitamin B_{12}. Twenty-five patients received vitamin B_{12} only. The patients who were treated with SAM-e had a significant improvement in liver function, including production of proteins, when compared with controls.

In another controlled trial also published in 1975, Dr. Cantoni and his team treated seventy inpatients who had chronic hepatitis as well as cirrhosis. The investigators were particularly interested in whether SAM-e would restore production of the protein albumin. For a twenty-day period, one group of patients received 15 mg IV of SAM-e twice a day, while a second group received a placebo IV. At the end of the study, the patients treated with SAM-e had their protein production restored.

A third study published in 1975 also analyzed protein production in patients with cirrhosis. Dr. Ideo conducted a double-blind study in which fifteen patients with cirrhosis received 15 mg of SAM-e four times a day (either IV or via injection into the muscle) for thirty days. A second group of fifteen patients was given the two precursors of SAM-e—methionine and ATP—for thirty days. Production of albumin returned to normal in the SAM-e treated patients but not in those who received the precursors for SAM-e. This study was particularly important because it indicated that the destruction of the SAM-e enzyme, SAM-e synthetase, is a key factor in cirrhosis. If that was not true, then giving the two precursors for SAM-e would have created enough SAM-e to restore protein production, as the SAM-e supplement did.

A subsequent study by a research team in Madrid, Spain, also looked at the role of SAM-e synthetase in cirrhosis. The investigators measured the activity of SAM-e synthetase in liver biopsies they obtained from seventeen healthy people and twenty-six individuals with cirrhosis. Both the activity of the SAM-e enzyme and the level of SAM-e itself were much lower in the individuals with cirrhosis when compared with the controls.

The Later Studies

In 1989, an Italian research team studied how SAM-e affects glutathione in people with cirrhosis. Glutathione is the potent antioxidant that is so important in the liver's ability to detoxify the blood. In this study, four groups of subjects were studied: nine patients with alcoholic cirrhosis received 1,500 mg of oral SAM-e daily for six months; seven patients with non-alcohol-related cirrhosis received the same SAM-e treatment; eight patients with alcoholic cirrhosis received placebo; and fifteen healthy individuals were the control group. At the beginning of the study, the glutathione levels were measured in all groups, and all patients with cirrhosis had significantly lower levels when compared with controls. The researchers found that both groups treated with SAM-e had a significant increase in the level of glutathione when compared with the placebo-treated group and concluded that "SAMe may exert an im-

portant role in reversing hepatic glutathione depletion in patients with liver disease."

Researchers in Mexico evaluated the benefit of
giving SAM-e by injection in the treatment of alcoholic cirrhosis. Forty-five patients were given either
SAM-e or placebo in the double-blind study. The placebo-controlled trial involved forty-five patients who
were treated for fifteen days. At the end of the study,
the researchers noted significantly favorable results in
the patients who had received SAM-e when compared
with those who had received placebo.

The deposit of excess cholesterol outside the liver
is another factor in liver disease. Dr. S. Rafique and
his colleagues at the Royal Free Hospital School of
Medicine in London, England, studied this problem
in 1992 in twenty-four patients with chronic liver disease. Thirteen of the patients were given daily oral
doses of SAM-e for two weeks, while eleven patients
received placebo. Ten of the SAM-e treated patients
had a decrease in the amount of cholesterol deposited
outside the liver, compared with four of the eleven
untreated patients. The researchers noted that "this
preliminary study is the first evidence . . . that a drug
can help to reverse the deposition of cholesterol in an
extrahepatic membrane."

There seems to be no end to the possible effects
SAM-e could have on liver disease, nor to the number
of different research teams in different countries interested in finding out what they are. In Japan in
1992, Dr. H. Kakimoto's team investigated whether
SAM-e changes the ability of materials to flow
through cell membranes (cell permeability), a function that would be beneficial in many different conditions, because it would improve the ability of

nutrients and other materials to enter the cells, while waste materials could leave more easily. Therefore they gave 600 mg IV of SAM-e for two weeks to sixteen patients with cirrhosis. Like the researchers in Rafique's study, Kakimoto measured the cholesterol levels in red blood cells and found that the levels decreased significantly after patients took SAM-e. They also noted that along with the decrease in cholesterol there was an increase in cell permeability of the red blood cell membranes. "These results suggest that SAM-e decreases [the cholesterol in red blood cell membranes] and thus improves membrane fluidity in chronic liver disease." The results of these two studies on cholesterol deposits and the increased fluidity of cell membranes support the concept that SAM-e improves the ability of fluids and other substances to be transported through cell membranes. This ability is particularly important in liver disease, because scarring of the liver prevents the proper flow of blood and other fluids through the liver for processing. SAM-e appears to help improve that situation.

An especially interesting review of more than one thousand cases of intrahepatic cholestasis—a condition in which bile flow is impaired in the liver—was done by Dr. M. Coltorti and his colleagues at the University of Naples, Italy. Intrahepatic cholestasis can be caused by cirrhosis, viral hepatitis, or drugs, as well as a few other less common causes. In Coltorti's 1990 review, the researchers noted that in controlled clinical trials in which patients with intrahepatic cholestasis were given 800 mg IV of SAM-e daily, they experienced a significant decrease in the biochemical factors that cause the disorder (e.g., bile salts and bilirubin) when compared with patients who had received a placebo. They also found that patients who had received both injectable (800 mg daily) and oral SAM-e (1,600 mg daily) had significant improvement in both the biochemical factors and in subjective symptoms, such as fatigue and general discomfort. Overall, Coltorti's team concluded that "SAMe represents an effective and safe approach to the management of intrahepatic cholestasis."

Coltorti's study was supported by a study in 1993, in which a trio of researchers reported in the *Postgraduate Medical Journal* that a patient with drug-induced (danazol) cholestasis recovered completely after receiving SAM-e intravenously for three weeks, followed by oral doses for six weeks

The liver needs bile salts to produce bile, and glutathione and taurine are needed to produce bile salts. This process is disrupted in people with cirrhosis. A study at La Sapienza University in Rome, Italy, examined the ability of oral SAM-e to help correct this problem. Ten patients with cirrhosis were examined before and two months after receiving 800 mg of oral SAM-e daily. The after-treatment levels of taurine and glutamic acid were higher (both substances are needed for liver functioning), and there was a significant increase in other factors favorable to the production of bile salts. The researchers concluded that "in the cirrhotic liver exogenous (supplemental) SAMe is partially metabolized to taurine, which is used for bile salt amidation."

SAM-e and Lead Poisoning

Two studies looked at the problem of "getting the lead out." Lead poisoning is a significant problem in the United States, especially among children. One in six children has toxic levels of lead in his or her blood, according to the Environmental Protection Agency. Lead in children can cause developmental problems, learning disabilities, and damage to nearly all the organs in the body. Most adults get contaminated from work-related sources.

One scientist who has studied lead poisoning is Dr. S. R. Paredes. He, along with several colleagues, conducted two studies of the therapeutic effect of SAM-e in lead intoxication in both mice and humans. In one study, mice with lead poisoning were given SAM-e via injection and orally for twenty days. Patients with chronic lead poisoning were given 12 mg/kg of SAM-e IV or an oral dose of 20 to 30 mg/kg. In both the mice and the patients, "the bulk of body lead burden was excreted in the feces, showing a peak within the first 24–48 hours." The amount excreted in the urine was low, and IV dosing resulted in a faster response than did oral dosing.

In a second study, SAM-e was given IV to two groups of mice: those with acute lead poisoning and those with chronic exposure to lead. The researchers found that the levels of lead dropped rapidly soon after treatment with SAM-e was initiated and that they reached safe levels after twenty to twenty-two days. Best results were obtained when they gave SAM-e over a period of twenty to twenty-two days. Dr. Paredes and his team concluded that "SAM therapy is beneficial in the treatment of lead intoxication."

The story of SAM-e and the liver is far from over. True, the studies conducted thus far have been promising. It appears that SAM-e helps restore much of the liver's ability to produce essential substances such as glutathione and proteins; that it improves the ability of fluids and materials to be transported through cell membranes; and that it is an effective treatment for intrahepatic cholestasis. But so far there have been no long-term clinical trials that can verify any of the findings made to this point. Many researchers have spent many years and much effort conducting the studies that have been reported thus far. Hopefully the research necessary to validate their work will not be long in coming.

NOTES

Angelico, M., Gandin, C., Nistri, A., et al. "Oral S-adenosyl-L-methionine (SAMe) administration enhances bile salt conjugation with taurine in patients with liver cirrhosis." *Scand J Clin Lab Invest* 54(6) (October 1994): 59–64.

Bray, G. P., Tredger, J. M., Williams, R. "Resolution of danazol-induced cholestasis with S-adenosylmethionine." *Postgraduate Medical Journal* 69(809) (March 1993): 237–239.

Coltorti, M., Bortolini, M., and DiPadova, C. "A review of the studies on the clinical use of S-adenosylmethionine (SAMe) for the symptomatic treatment of intrahepatic cholestasis." *Methods Find Exp Clin Pharmacol* 12(1) (January-February 1990): 69–78

Diaz, Belmont A., Dominguez, Henkel R., and Uribe, Ancira F. "Parenteral S-adenosylmethionine compared to placebo in the treatment of alcoholic liver disease." *Am Med Interna* 13(1) (January 1996): 9–15.

Duce, A. M., Ortiz, P., Cabrero, C., and Mato, J. M. "S-adenosyl-L-methionine synthetase and phospholipid methyltransferase are inhibited in human cirrhosis." *Metabolismo, Nutricion y Homonas,* 8(1) (January-February 1988): 65–68.

Environmental Protection Agency. Office of Pollution Prevention and Toxics. EPA Doc. #800-B-92-002.

Kakimoto, H., Kawata, S., Imai, Y., et al. "Changes in lipid composition of erythrocyte membranes with administration of S-adenosyl-L-methioine in chronic liver disease." *Gastroenterol Jpn* 24(4) (May 1989): 407–415.

Labo, G., Miglio, F. G., et al. "Double-blind polycentric study of the action of S-adenosyl-L-methionine in hepatic cirrhosis." *Minerva Medica* 66(33) (2 May 1975): 1590–1594.

Loguercio, C., Nardi, G., Argenzio, F., et al. *Alcohol Alcohol* 29(5) (September 1994): 597–604.

Paredes, S. R., Fukuda, H., Kozicki, P. A., et al. "S-adenosyl-L-methionine and lead intoxication: Its therapeutic effect varying the route of administration." *Ecotoxicol Environ Saf* 12(3) (December 1986): 252–260.

Paredes, S. R., Kozicki, P. A., and Battle, A. M. "S-adenosyl-L-methionine: A counter to lead intoxication?" *Comp Biochem Physiol* 82(4) (1985): 751–757.

Pisi, E., Marchesini, G. "Mechanisms and consequences of the impaired transsulphuration pathway in liver disease. Part II. Clinical consequences and potential for pharmacological intervention in cirrhosis." *Drugs* 40(Suppl 3) (1990): 65–72.

Rafique, S., Guardascione, M., Osman, E., et al. "Reversal of extrahepatic membrane cholesterol deposition in patients with chronic liver diseases by S-adenosyl-L-methionine." *UK Clin Sci* 83(3) (September 1992): 353–336.

Vendemiale, G., Altomare, E., Trizio, T., et al. "Effects of oral S-adenosyl-L-methionine on hepatic glutathione in patients with liver disease." *Scand J Gastroenterol* 24(4)(May 1989): 407–415.

8

More Reasons for Hope: SAM-e and Other Medical Conditions

Thus far, you have read about how SAM-e works in the treatment of depression, osteoarthritis, fibromyalgia, and liver disease. In all of theses cases, SAM-e has been found to be of benefit, to varying degrees, and its use has been supported by clinical studies.

Because SAM-e is a methyl donor and participates in so many different methylation processes, it has far-reaching effects in the body, as research has borne out. This is a reason why scientists keep finding new potential uses for SAM-e while investigating its benefits in other areas. Investigators are still discovering the secrets of SAM-e, and probably will be doing so for many years to come.

Many investigators have traveled the research road from depression to osteoarthritis to fibromyalgia to cirrhosis and back again, riding its many twists and turns. Some of those scientists have taken the side avenues and explored the "lesser traveled roads" to uncover what they can about how SAM-e works in other areas, including heart disease, migraine, attention deficit disorder, Alzheimer's disease, Parkinson's disease, and aging in general. The research is not as plentiful, nor as extensive, but it is still in its infancy. In some cases it is very preliminary, but in all cases it is also exciting, because of the potential SAM-e has in either directly treating or complementing the treatment of these conditions.

This chapter looks at the research that has been done in these other areas, as well as background information on each of the conditions. Even as you are reading this chapter, new discoveries are being made about the ways SAM-e may be a part of the solution to some of the disorders that affect our health the most.

Cardiovascular Disease

The number one cause of death in the United States is cardiovascular disease (CVD), or heart disease. More people die of heart disease than cancer, infectious diseases, and murder combined. According to the American Heart Association, cardiovascular disease claimed the lives of 959,227 people in the United States in 1996. This is 41.4 percent of all deaths. Since 1900, CVD has been the number one killer in the United States for every year except 1918. More than 2,600 Americans die each day of CVD, which is an average of one death every thirty-three seconds.

The term "heart disease" refers to any disease that affects the heart's blood vessels, or the coronary arteries. The most common condition that blocks the coronary arteries is atherosclerosis, which is an accumulation of plaque (a combination of cholesterol, fatty substances, and cellular waste) in the blood vessels. A heart attack is the result of a blockage of the arteries that supply the heart; when an artery in the brain is blocked, the result is a stroke.

Types of Cardiovascular Disease

Cardiovascular disease includes peripheral artery disease, coronary artery disease, myocardial infarction (commonly known as heart attack), stroke, aneurysm, thromboembolism, extracranial carotid artery, and stenosis. Generally what occurs in heart disease is that when an artery or vein becomes diseased, the inner walls become thick. The cells in the vessel lining combine with other materials and form atherosclerotic plaque. Plaque consists of cholesterol, fats, and cellular debris and is considered to be the first sign of cardiovascular disease. Plaque evolves over time, becomes thicker, and distorts the artery wall with a condition called atheroma. When an atheroma blocks the flow of blood to the heart, it is commonly called a heart attack. When the blockage occurs in a blood vessel going to the brain, it is a stroke. Partial blockage in the chest is known as angina, and ruptures of the arteries or veins are called aneurysms.

The major risk factors for heart disease are normally recognized to be smoking, high blood levels of cholesterol, high blood pressure, diabetes, and lack of physical activity. Other risk factors, although classified as "minor," often have been found to be more significant than the major risk factors, according to Michael Murray, N.D., and a growing number of physicians and scientists. A few of these minor risk factors include low levels of essential fatty acids and antioxidants; low levels of magnesium and potassium;

elevated levels of homocysteine; and a type A personality (a person who has a high level of stress in his or her life).

The Buzz about Homocysteine

Recently, there has been much interest in one of these minor risk factors—elevated levels of the amino acid homocysteine. Many researchers believe this factor is critically important to heart disease, even more important than high cholesterol levels. Homocysteine is a natural byproduct of the metabolism of methionine. If the right elements are present, homocysteine eventually converts into cysteine and other beneficial substances. Some of it also converts back to methionine. But if the elements are not present, homocysteine accumulates and reaches toxic levels.

Studies show that homocysteine irritates the inner lining of arteries and veins and causes plaque buildup. According to a recent article in the medical journal *Circulation*, high levels of homocysteine impair the ability of blood vessels to dilate (expand). Some researchers believe that rather than give people drugs that dilate their blood vessels, it may be possible to achieve the same result by taking nutrients or supplements that lower homocysteine levels. SAM-e is one such supplement.

SAM-e, as you will recall, is the result of the combining of methionine and ATP. The role of SAM-e in homocysteine goes like this (the entire process is explained in more detail in Chapter 2): methionine and ATP combine to form SAM-e; a SAM-e molecule gives up a methyl group, and the remaining molecule is converted to a substance called S-adenosyl-homocysteine. This new molecule then donates its sulfur

("S") to other molecules, and in the process it also loses its "adenosyl" (adenosine) portion. What remains is homocysteine. Homocysteine than breaks down and releases the amino acid cysteine, or the homocysteine can be converted back into methionine. In a healthy system, this process of breakdown and conversion goes smoothly, and just enough homocysteine is produced (a little is good; a lot is bad).

Just how bad high levels of homocysteine can be was demonstrated in a large study published in the *Journal of the American Medical Association* in 1992. Dr. M. J. Stampfer and his colleagues accumulated data on nearly fifteen thousand healthy male physicians who were part of the Physicians' Health Study. Over a five-year period, the researchers found a greater than threefold increase in the risk of heart attack among physicians who had an elevated homocysteine level. These high levels are dangerous, researchers believe, because homocysteine appears to promote accumulation of plaque in the arteries by damaging the walls of the arteries and reducing their integrity, causing the condition known as atherosclerosis. (More on the role of homocysteine in heart disease is discussed below under "Homocysteine and Cardiovascular Disease.")

How do homocysteine levels become elevated? Although some cases of high homocysteine may be related to an inherited abnormality, most are associated with improper diet. In Chapter 1, it was mentioned that the nutrients folic acid, vitamin B_{12}, and vitamin B_6 should be taken with SAM-e supplements (also see the Graham study below). These nutrients help SAM-e perform its many functions, one of which is the methylation process that has homocysteine as a byproduct. People who are deficient in these three B vitamins also have high levels of homocysteine. Therefore, sufficient levels of these nutrients help ensure that SAM-e will function optimally and that homocysteine levels will remain low.

Cholesterol and Cardiovascular Disease

Much has been written about the role of high cholesterol levels in cardiovascular disease. Many researchers are now realizing that although cholesterol levels do correlate with heart disease, they are not the cause of it. This does not mean, however, that you should not maintain a low cholesterol level. High cholesterol *is* still a risk factor for heart disease.

Evidence that elevated cholesterol is not as critical in heart disease as once believed can be found in several recent studies. One of the largest studies was reported by Dr. M. J. Stampfer in the *Journal of the American Medical Association* (discussed above) and supports the idea that high homocysteine levels precede coronary artery disease. In a review article which included the results from 25,968 individuals in thirty-four studies, the reviewers found that even though reducing cholesterol levels by 10 percent or more did reduce the risk of dying of a heart attack in middle-aged men, they were not able to reach any conclusions about the elderly or women.

Homocysteine and Cardiovascular Disease

Dozens of studies have been done that present evidence that elevated homocysteine levels are a major factor in cardiovascular disease. One such study was conducted in 1997 by Dr. I. M. Graham and his associates, of Trinity College in Dublin, Ireland. The investigative team evaluated 750 people who had vascular disease and compared them with 800 healthy controls. The participants were drawn from nineteen centers in nine European countries. The homocysteine levels, as well as the levels of folic acid, vitamin B_{12}, vitamin B_6, and cholesterol of all patients and controls were noted both before and after they received a dose of methionine. After the dose of methionine, the investigators noted several things. Overall, they reported that an increase in homocysteine level was "an independent risk of vascular disease similar to that of smoking or hyperlipidemia [high fat content in the blood]. It powerfully increases the risk associated with smoking and hypertension." They also noted that the higher the homocysteine level, the lower were the levels of folic acid and vitamins B_6 and B_{12}, a relationship that has also been associated with heart disease and depression. Thus it was no surprise that people in the study who took vitamin supplements had a substantially lower risk of vascular disease.

In Hong Kong, Dr. K. S. Woo and his colleagues identified high homocysteine levels as a risk factor for arterial endothelial dysfunction—a fancy name for a condition that precedes atherosclerosis. Using two special tests, they evaluated the dilation in the brachial artery in fourteen healthy individuals who had high homocysteine levels and fourteen controls with low homocysteine levels. Those with high homocysteine levels had significantly lower dilation than the subjects with low homocysteine. Therefore even though the individuals with high homocysteine appeared to be healthy and they had no obvious symptoms of heart disease, the poor dilation caused by high homocysteine placed them at risk for cardiovascular disease.

Actually, being asymptomatic (having no obvious symptoms) for heart disease is not uncommon at all. A study done in the 1970s (the results of which still hold true) revealed that more than 50 percent of the people who were brought into the coronary care units in the hospital for their first myocardial infarction did not have any of the risk factors for heart disease. The presence of high homocysteine is a silent risk factor.

In the Netherlands, investigators examined whether high homocysteine levels are a risk factor for vascular disease among elderly men. Dr. C. D. Stehouwer and his colleagues followed 878 men for ten years to determine if those with high homocysteine levels (31 percent of the total group) had a significantly greater risk of dying of coronary heart disease. They found that a high homocysteine level is a "strong predictive factor for fatal cerebrovascular disease in men without hypertension [high blood pressure] but less so for coronary heart disease." Thus an increased risk of dying of stroke has also been added to the list of cardiovascular conditions associated with a high level of homocysteine.

In a study conducted at the USDA Human Nutrition Research Center on Aging in Boston, Massachusetts, investigators looked at the homocysteine levels and the death rate from all causes and from cardiovascular disease among elderly individuals who were in the Framingham Heart Study. (The Framingham Heart Study, begun in 1947, is a landmark study that evaluated the development of coronary heart disease in a large—more than five thousand people—adult population. The study continues today.) The current study included 1,933 men and women who were examined between 1979 and 1982 and then reexamined through 1992. The researchers concluded that high levels of homocysteine "are independently associated with increased rates of all-cause and CVD mortality in the elderly."

SAM-e, Homocysteine, and Cardiovascular Disease

The idea that SAM-e might be a key player on the scene with high homocysteine and cardiovascular disease was presented by Dr. Franziska Loehrer and his colleagues at University Hospital in Switzerland. Dr. Loehrer and his team had noted that "high levels of homocysteine values in patients after an interval (of approximately) one year supports the idea that this parameter [homocysteine] plays a role in the disease process." But they went one step further. After explaining that cholesterol and triglyceride levels (fatty acids in the blood) did not have a statistically significant correlation with coronary artery disease, they stated that the factors that were significant included elevated homocysteine levels, low SAM-e levels, age, and body mass. They then went on to report their research.

One of the studies was published in 1996 in *Arteriosclerosis, Thrombosis and Vascular Biology,* in which Loehrer and his colleagues explained the link among high homocysteine levels, coronary artery disease (CAD), and SAM-e. These Swiss researchers studied seventy coronary artery disease patients and controls and for the first time found an association between high levels of homocysteine and an enzyme (5-methyltetrahydrofolate; also referred to as folate) that is responsible for homocysteine metabolism—the conversion of homocysteine into methionine. SAM-e plays a critical role in preventing the breakdown of this enzyme and in the process that allows the conversion of homocysteine to take place. The researchers also found that the patients with coronary artery disease had low levels of SAM-e. Ultimately the study indicated that there is a correlation between high homocysteine levels, low SAM-e levels, and heart disease, and "that SAM-e might be a protective factor against the development of CAD."

Loerher and his team continued with other studies. In the same year they looked at the effect of giving methionine (one of SAM-e's components) on homocysteine, SAM-e, and folate. They gave methionine to twelve healthy individuals over a twenty-four-hour period. They observed the expected rise in homocysteine and SAM-e levels, and a decrease in folate. They concluded that "a change in either homocysteine or S-adenosylmethionine may cause a reduction in 5-methyltetrahydrofolate," and that this information "must be considered in evaluating the relationship between folate and homocysteine in vascular disease" and the role SAM-e plays in this scenario.

Then in 1997, Dr. Loerher again looked at the question of SAM-e and homocysteine metabolism, in fourteen people without heart disease. His team administered 400 mg of oral SAM-e to the patients and noted its effect on 5-methyltetrahydrofolate, S-adenosylhomocysteine (a precursor to homocysteine), homocysteine, and methionine. They observed an increase in the levels of 5-methyltetrahydrofolate, which is a key element in homocysteine metabolism. The researchers concluded that their finding "should be considered in homocysteine-lowering strategies for the prevention of vascular disease."

Many questions remain about the role of SAM-e in the prevention or treatment of cardiovascular disease. Thus far there is good evidence that it is a factor. As the role of elevated levels of homocysteine in this disease process becomes more defined and more studies are done with SAM-e and heart disease, the way may be made clear for a definitive role of SAM-e in the prevention and treatment of cardiovascular disease.

Parkinson's Disease

Parkinson's disease is a chronic, slowly progressive disease in which a group of brain cells that control body movement die. It is also a disease in which SAM-e levels are low, hence the interest in the role SAM-e supplementation may make in this disorder.

Parkinson's disease, which can strike people of any age but which usually first appears at age forty or older, affects more than 500,000 people in the United States. Its primary symptoms are rigidity, which is defined as stiffness in the leg, arm, or neck when moved; resting tremor, which is a shaking motion when sitting quietly; bradykinesia, which is slowness when initiating a movement; and loss of postural reflexes, which affects balance and coordination. Bradykinesia causes changes in speech, the characteristic shuffling gait seen in many Parkinson's disease patients, and decreased facial expression. About 60 percent of Parkinson's disease patients have resting tremor as a symptom. Other common symptoms include depression, memory and sleep problems, difficulty chewing or swallowing, urinary or bowel problems, and low blood pressure when standing. Symptoms are not the same for every person with the disease, and they often progress slowly over decades.

Because there are no tests to detect Parkinson's disease, doctors must rely on a patient's history and a careful examination. The disease is caused by the gradual death of cells that lie deep within the brain in an area called the substantial nigra, where the neurotransmitter dopamine is produced. (Dopamine also has a role in depression; see Chapter 3.) Dopamine travels throughout the brain, including a portion called the striatum, which coordinates various brain circuits. When there is a deficiency of dopamine in the striatum, symptoms of Parkinson's develop. As the disease develops, cells in other portions of the brain and nervous system also degenerate and die. The reason the cells die is not known. One theory is that exposure to toxins triggers cell death; others propose genetic factors and accelerated aging.

Treatment of Parkinson's disease consists of medications that treat symptoms only. These include levodopa and carbidopa, which help restore the brain's low supply of dopamine. These drugs are often used together. Other drugs called dopamine agonists (eg., bromocriptine and pergolide), which mimic the action of dopamine, can delay the need to prescribe levodopa, and have fewer side effects. Side effects of medications include depression, nausea, low blood pressure, vomiting, involuntary movements, and restlessness. No drug or treatment has been found thus far that can cure or stop the progression of Parkinson's disease. Surgical procedures, including the transplantation of brain cells from the substantia nigra of a human fetus or cloned animal brain cells into the brains of Parkinson's disease patients, are controversial and still under investigation. Procedures known as pallidotomy and thalamotomy, which create a permanent lesion in the brain, can reduce some symptoms for selected patients.

SAM-e and Parkinson's Disease

In 1990, a double-blind, cross-over study looked at the effect of SAM-e in patients with Parkinson's disease. Because levodopa depletes the levels of both SAM-e and serotonin, people with this disease are very susceptible to depression. In fact, the incidence of depression in these patients is about 46 percent. Yet after only two weeks of treatment with SAM-e, 72 percent of the Parkinson's disease patients in a study conducted by Dr. P. B. Carrieri had a definite improvement in depressive symptoms, compared with only 30 percent of patients who took a placebo. Most importantly, SAM-e did not affect levodopa treatment.

The idea that SAM-e may help people with Parkinson's disease is intriguing, especially since the current treatment is not effective indefinitely. Research shows that levodopa loses its effectiveness after about four years. Several studies suggest that the reason for this is that levodopa depletes serotonin. In a study in *Neuroscience Letters* in 1993, the researchers showed that Parkinson's disease patients have significantly lower levels of dopamine and serotonin in their cerebral spinal fluid, and that patients who take levodopa have even lower serotonin levels than untreated patients. Other research shows that serotonin raises dopamine levels and that dopamine lowers serotonin levels in certain parts of the brain. This sort of checks-

and-balances act appears to ensure that neither neurotransmitter reaches levels that are too high. This relationship must also be kept in mind when researchers explore the possible role of SAM-e in Parkinson's disease, because SAM-e has an effect on both of these neurotransmitters.

It may be easy to assume that because Parkinson's disease patients have low levels of dopamine, serotonin, and SAM-e, people with the disease should take a supplement of SAM-e to raise those levels. However, Parkinson's disease is a situation in which the obvious may not be what it seems. Parkinson's disease is an extremely complex disorder. Although the patients in Dr. Carrieri's study experienced relief of depressive symptoms, the question remains whether continuing SAM-e treatments would be safe and effective against other symptoms of Parkinson's disease.

Dr. Grazi raises the issue in *The European Arthritis and Depression Breakthrough! SAMe,* when he suggests that the depletion of SAM-e by dopamine may be a safeguard by the body. During the methylation process in which dopamine levels are reduced, the level of another neurotransmitter, acetylcholine, increases. This shift in neurotransmitter levels causes the tremors and impaired movement associated with Parkinson's disease. Dr. Grazi suggests that because SAM-e is a methyl donor, the body may intentionally reduce the level of SAM-e in an attempt to slow the methylation process and thus interfere with the effects of the disease. In such a scenario, taking SAM-e would be detrimental to people with Parkinson's disease.

The question of whether to take SAM-e for Parkinson's disease—for the depression or any other symptoms—remains unanswered, but it is certainly fertile ground for research. Chances are more studies will be forthcoming in this area in the near future. If you have Parkinson's disease and want to consider taking SAM-e, discuss it with your physician.

SAM-e, Alzheimer's Disease, and Aging

It's a fact: as we age, levels of many substances in the body decline; processes that worked without a hitch when we were younger begin to slow down or break down; and mental function begins to decrease. In addition to a decline in the level of SAM-e and the antioxidant glutathione, the aging body does not assimilate or use certain nutrients (e.g., vitamin D, vitamin B_{12}) as well as a younger body. Metabolism is not as efficient, the immune system's ability to respond to invading organisms is reduced, and short-term memory loss has been shown to begin as early as age forty.

No one factor is responsible for these changes, nor are they all an inevitable part of growing older. Some of the things that contribute to aging include drugs that many people take for physical ailments, which have the side effects of creating confusion, memory loss, and depression; the accumulation of toxins in the body over the years; a deficiency of antioxidants to fight free-radical damage (especially the "age buster" glutathione, see Chapter 2); the effects of years and years of repetitive movements that cause joint degeneration as in osteoarthritis; poor nutrition; smoking; drug and alcohol abuse; and many other factors.

SAM-e and Aging

SAM-e's role in preventing aging is tied with its extensive involvement in so many bodily processes. SAM-e is essential for the synthesis of neurotransmitters, DNA, and phosphatidylcholine, a substance that keeps the cell membranes flexible and allows cells to more efficiently pass nutrients and wastes through its walls (a process called cell permeability). It may help prevent heart disease and liver disease and the crippling effects of osteoarthritis.

SAM-e is also the precursor for the powerful antioxidants glutathione and cysteine, which fight free radicals that cause aging. This may be the most important job SAM-e does in the area of aging. By providing the brain and the rest of the body with a healthy supply of antioxidants, SAM-e helps prevent the damaging effects of free radicals. Free radicals have a special affinity for fat, and the brain is more than 50 percent fat. Thus the potential for damage to and destruction of the nerve cells and neurotransmitters in the brain increases as we age. One of the diseases that is characterized by a decreased amount of neurotransmitters is Alzheimer's disease.

Alzheimer's Disease

Alzheimer's disease is a degenerative brain disorder in which people experience a progressive deterioration of their mental functions and of their memory. This condition is also known as dementia. About 15 percent of people in the United States suffer from some degree of dementia.

When autopsies are performed on people who had dementia, between 50 and 60 percent of them turn out to have had Alzheimer's. Unfortunately, there are no definitive tests to diagnose Alzheimer's disease while a person is still alive, although magnetic resonance imaging and computed tomography scans can be used to detect atrophy in the brain, which suggests Alzheimer's. Autopsy studies show that the brain undergoes changes in which deposits of cellular debris and proteins accumulate and eventually destroy brain cells, especially in the areas associated with mental functioning and memory. A decline in the levels of the neurotransmitter acetylcholine is associated with memory loss.

SAM-e and Alzheimer's Disease

Researchers have found that people with Alzheimer's disease have lower levels of SAM-e than those without the disease; however the reason for this is not known. Studies in rodents have provided us with most of the information we have on SAM-e and Alzheimer's disease. Some of those studies suggest SAM-e levels are low because production decreases with age, while others support the idea that the body depletes its supply of SAM-e more rapidly as we age.

At least one other theory appears to have been discarded. The idea that the decline in SAM-e may be due to the degeneration of the brain itself is not believed to be true, at least according to Dr. Morrison at the University of Toronto, Ontario, Canada. He and his colleagues examined the brains of eleven patients with Alzheimer's disease who had died. They found greatly decreased levels of SAM-e when compared with the brains of ten patients who had had Parkinson's disease. This finding suggested to them that low levels of SAM-e "are not simply a consequence of a chronic neurodegenerative condition."

However, researchers still do not know why SAM-e levels are low in people with Alzheimer's disease.

The value of SAM-e for treatment of people with Alzheimer's disease may lie with its ability to increase the flow of nutrients and other materials in and out of the cells. This feature of SAM-e was discussed in detail in Chapter 2, in the section about SAM-e and methylation of phospholipids. It's already been shown that improved cell permeability is an important part of SAM-e's ability to help those with liver disease. Therefore it is proposed that because phospholipid methylation declines as we age, supplementation with SAM-e will increase cell permeability and thus the flow of neurotransmitters, which will help people with Alzheimer's disease.

This idea was tested in 1988 when researchers at McLean Hospital in Belmont, Massachusetts, administered SAM-e to patients with Alzheimer's disease to see if the increase in cell membrane permeability resulted in any benefit. Although the investigators noted a significant increase in membrane fluidity, they did not note any improvement or worsening of symptoms. They concluded that "while SAM-e may be useful for other conditions associated with altered membrane fluidity (such as normal aging), changing membrane fluidity per se is not likely to lead to marked changes in symptoms of AD [Alzheimer's disease]."

Studies with SAM-e in Other Medical Conditions

Scattered throughout the medical literature are studies of the effectiveness of SAM-e in various, isolated medical conditions. They are discussed briefly here because in several other cases with SAM-e, useful results have been born of serendipitous findings. One of the studies noted here may eventually result in a common, effective treatment, more research, or another link in the secret of SAM-e.

Skin Disease

A study published in the March 1987 issue of the *British Journal of Dermatology* reported on the use of SAM-e and a drug called chloroquine in the treatment of infantile porphyria cutanea tarda, a rare inherited disorder in which there is a disturbance of the metabolism of the breakdown products of the hemoglobin. Two children, a seven-year-old girl and a twelve-year-old boy, had been diagnosed with porphyria cutanea tarda. They were treated with a combination of oral SAM-e and low-dose oral chloroquine. Within three months, both children attained complete clinical and biochemical recovery without experiencing any side effects. The investigators proposed that the treatment used in their study (oral SAM-e at 12 mg/kg daily for three weeks and then oral chloroquine 100 mg twice weekly for up to 150 days or until improvement is reached) is the treatment of choice. They concluded that "this combined therapy appears to be safe, simple, effective and very convenient for both patients and physicians."

Migraine

Those who suffer with migraine know the magnitude of the pain: a pounding or throbbing sharp pain on one side of the head, sometimes preceded by an "aura"—visual disturbances, nausea, and gastrointestinal upset. About 80 percent of people who experience migraines get the "common" type, which typically lasts one to three days and generally is not accompanied by aura. The remaining 20 percent get either the "classic" migraine, which usually lasts two to six hours and is characterized by aura, or what is known as "complicated" migraine, which has varying and unpredictable signs.

Current drug methods to treat migraine are not very effective. One thing that complicates any attempt to treat migraine is that researchers are not sure exactly what causes it. Some studies indicate that people with migraine have an abnormality in the way the blood vessels constrict and dilate (expand). Others suggest that a compound called substance P is released into the bloodstream. Substance P triggers pain and also causes the blood vessels to dilate. Still others point to a deficiency of serotonin. A low level of the neurotransmitter serotonin causes a decrease in the pain threshold. More than thirty-five years of research support this theory as a cause of migraine.

Only one study has been published about the use of SAM-e in the treatment of migraine. The investigators found that long-term use of SAM-e provided significant pain relief and that the benefit built up gradually over time. They attributed the positive results to SAM-e's ability to raise serotonin levels. Elevated serotonin levels are associated with pain relief.

Attention Deficit Hyperactivity Disorder

A preliminary study conducted in 1990 suggested a relationship between SAM-e and attention deficit hyperactivity disorder (ADHD) in children and adults. Currently, treatment for ADHD consists of medications such as methylphenidate (Ritalin) that increase the activity of dopamine and norepinephrine. Ritalin has many side effects, has a high potential for abuse, and is controversial, especially for use in children. The researchers conducted several trials at the Neuropsychiatric Institute, University of California, Los Angeles, among adults with ADHD: a four-week open trial and a nine-week, double-blind, placebo-controlled cross-over trial. Preliminary data indicated that six of the eight patients (75 percent) had moderate to marked improvement when taking oral doses of SAM-e. The two patients who did not respond also had not responded to Ritalin treatment.

Gastrointestinal Protection

In the *American Journal of Medicine*, Oscar M. Laudanno, M.D., reported that in a comparison study conducted on rats, both SAM-e and misoprostol (a drug used to help prevent gastric injury caused by the use of nonsteroidal anti-inflammatory drugs, or NSAIDs) were significantly effective in reducing injury to the stomach lining. This benefit of SAM-e is like getting a bonus, especially for people who have been taking NSAIDs, which are known to cause gastrointestinal distress. Since Dr. Laudanno's study (in 1987), the gastrointestinal protective ability of SAM-e has been well noted and has proved especially beneficial for people who were once taking NSAIDs for osteoarthritis and who are now taking SAM-e.

Dozens of other studies have been conducted on animals on the effects of SAM-e on different processes and substances in the body. These include experiments to evaluate SAM-e's effect on pain and inflammation in general, and on cancer cells, especially in the liver. Perhaps these investigations will uncover information that spurs researchers on to do studies in humans and introduce new uses for SAM-e, or that will further support the research done thus far.

NOTES

Battle, A. M., Stella, A. M., De Kaminsky, A. R., et al. "Two cases of infantile porphyria cutanea tarda: Successful treatment with oral S-adenosyl-L-methionine and low-dose oral chloroquine." *British Journal of Dermatology* 116(3) (March 1987): 407–415.

Bostom, A. G., Silbershatz, H., Rosenberg, I. H., et al. "Nonfasting plasma total homocysteine levels and all-cause and cardiovascular disease mortality in elderly Framingham men and women." *Archives of Internal Medicine* 159(10) (24 May 1999): 1077–1080.

Cheng, H., Gomes-Trolin, C., Aquilonius, S. M., et al. "Levels of L-methionine S-adenosyltransferase activity in erythrocytes and concentrations of S-adenosylmethionine and S-adenosylhomocysteine in whole blood of patients with Parkinson's disease." *Exp Neurol* 145(2 Pt 1) (June 1997): 580–585.

Clarke, R., et al. "Hyperhomocysteinemia: An independent risk factor for vascular disease." *New England Journal of Medicine* 324 (1991): 1149–1155.

Cohen, B. M., Satlin, A., and Zubenko, G. S. "S-adenosyl-L-methionine in the treatment of Alzheimer's disease." *J Clin Psychopharmacol* 8(1) (1988): 43–47.

Crowell, B. G., Benson, R., Shockley, D., and Charlton, C. G. "S-adenosyl-L-methionine decreases motor activity in the rat: Similarity to Parkinson's disease-like symptoms." *Behavioral & Neural Biology* 59(3) (May 1993): 186–193.

Gatto, G., Caleri, D., Michelacci, S., and Sicuteri, F. "Analgesizing effect of a methyl donor (S–adenosylmethionine) in migraine. An open clinical trial." *Int J Clin Pharmacolo Res* 6(1) (1986) 15–17.

Glueck, C. J., et al. "Evidence that homocysteine is an independent risk factor for atherosclerosis in hyperlipidemic patients." *Am J Cariol* 7 (1995): 132-136.

Graham, I. M., Daly, L. E., Refsum, H. M., Robinson, K., et al. "Homocysteine, vascular disease." *JAMA* 277(22) (11 June 1997): 1775–1781.

Jenkins, C. D. "Psychologic and social precursors of coronary disease." *New England Journal of Medicine* 284 (5) (4 February 1971): 244–255.

Laudanno, Oscar M. "Cytoprotective effect of S-adenosylmethionine compared with that of misoprostol against ethanol, aspirin, and stress-induced gastric damage." *American Journal of Medicine* 83 (Suppl 5A) (20 November 1987).

Loehrer, F. M., Angst, C. P., Haefeli, W. E., et al. "Low whole-blood S-adenosylmethionine and correlation between 5-methyltetrahydrofolate and homocysteine in coronary artery disease." *Arterioscler Thromb Vasc Biol* 16(6) (June 1996): 727–733.

Loehrer, F. M., Haefeli, W. E., Angst, C. P., et al. "Effect of methionine loading on 5-methyltetrahydrofolate, S-adenosylmethionine and S-adenosylhomocysteine in plasma of healthy humans." *Clin Sci (Colch)* 91(1) (July 1996): 79–86.

Loehrer, F. M., Schwab, R., Angst, C. P., et al. "Influence of oral S-adenosylmethionine on plasma 5-methyltetrahydrofolate, S-adenosylhomocysteine, homocysteine and methionine in healthy humans." *Pharmacol Exp Ther* 282(2) (August 1997): 845–850.

Marcholi, R., Marfisi, R. M., Carinci, F., and Tognoni, G. "Meta-analysis, clinical trials, and transferability of research results into practice. The case of cholesterol-lowering intervention in the secondary prevention of coronary heart disease." *Archives of Internal Medicine* 156 (1996): 1158–1172.

Morrison, L. D., Smith, D. D., and Kish, S. J. "Brain S-adenosylmethionine levels are severely decreased in Alzheimer's disease." *Journal of Neurochemistry* 67(3) (September 1996): 1328–1331.

Shekim, W. O., Antun, F., Hanna, G. L., et al. "S-adenosyl-L-methionine (SAM-e) in adults with ADHD, RS: Preliminary results from an open trial." *Psychopharmacol Bull* 26(2) (1990): 249–253.

Stampfer, M. J., Malinow, M. R., Willett, W. C., et al. 1992. "A prospective study of plasma homocysteine and risk of myocardial infarction in U.S. physicians." *Journal of the American Medical Association* 268: 877–881.

Stehouwer, C. D., Weijenberg, M. P., van den Berg, M., et al. "Serum homocysteine and risk of coronary heart disease and cerebrovascular disease in elderly men: A 10-year follow-up." *Arterioscler Thromb Vasc Biol* 18(12) (December 1998): 1895–1901.

Stramentinoli, G., Gaulano, M., Catto, E., and Algeri, S. "Tissue levels of S-adenosylmethionine in aging rats." *Journal of Gerontology* 32(4) (July 1977): 392–394.

Surtees, R., and Hyland, K. "Cerebrospinal fluid concentrations of S-adenosylmethionine, methionine, and 5-methyltetrahydrofolate in a reference population: Cerebrospinal fluid S-adenosylmethionine declines with age in humans." *Biochemical Medicine & Metabolic Biology* 44(2) (October 1990): 192–199.

Woo, K. S., Chook, P., Lolin, Y. I., et al. "Hyperhomocyst(e)inemia is a risk factor for arterial endothelial dysfunction in humans." *Circulation* 96(8) (21 October 1997): 2542.

9

Everything You Need to Know about Taking SAM-e

If you've made it to this chapter, you're truly curious about SAM-e and what it can do for you or someone you care about. You've read all the information in the previous chapters, or perhaps just the chapters that apply to you. Now you're ready to try it, or think you are. But first you have some questions, and well you should. No one should ever start to take a supplement without first finding out all they can about it.

Another recommendation you often hear is to consult with your physician before beginning a supplement program or taking a new medication. Sure, you say, everyone always tells you that, but is it really necessary? If you just want to start taking vitamin C or a multivitamin-mineral, conferring with your physician probably isn't necessary for the majority of people. If you feel a cold coming on, taking some echinacea and zinc doesn't require that you run to your physician for approval.

But SAM-e can be different. Not because it is dangerous, which it does not appear to be in any of the studies done over the last twenty-five-plus years, but because it can make a significant difference in your health and how your body assimilates and utilizes other nutrients and substances. And because self-diagnosis is not always a good idea, especially if you are in poor health, have a chronic medical condition, or are taking medication. In such cases, it is particularly recommended that you consult with your physician before starting any treatment with SAM-e.

According to a 5 July 1999 article in *Newsweek*, some experts are expressing real concern over the use of SAM-e for depression, saying that "the nation is embarking on a large, uncontrolled experiment in which consumers are the guinea pigs." Much of this feeling stems from the fear that depressed individuals might suffer adverse effects if they stop taking their conventional medication, switch to SAM-e, and do not get relief. To date, however, there have been no reports of any major side effects or deaths from taking SAM-e.

Some people are already taking SAM-e to treat mild depression or the "blues," and they have not sought medical guidance. As long as SAM-e is working for them and they aren't experiencing any ill effects that require medical attention, they probably don't need to see a doctor. But if you are experiencing moderate or severe depression; if you have moderate or severe osteoarthritis and are taking medications; or if you have liver disease, by all means, consult a physician before starting SAM-e. (As mentioned earlier, if you have bipolar depression the SAM-e supplement is not for you.) You can take this book with you to your health-care provider so the two of you can discuss your treatment options.

This chapter answers questions about how to take SAM-e so you can give it a fair chance to see if it works for you. It covers how to take SAM-e for the main medical conditions discussed in the previous chapters, as well as how to take it with other supplements and drugs, and its side effects. It also explains how to optimize the benefits of your SAM-e supplement

Meet SAM-e the Supplement

No supplement or drug will do what it reportedly can do if it is somehow damaged, contaminated, or otherwise rendered ineffective. Some supplements must be kept refrigerated to maintain their potency. Several forms of the beneficial bacteria supplement called acidophilus, for example, must be kept cool or the organisms will die and the supplement will not work. It is also recommended that you keep oil-based vitamin E capsules in the refrigerator to prevent them from going rancid. Most supplement containers have a statement that says "Store in a cool dry place" because both heat and moisture can deplete supplements of their potency.

SAM-e is a highly unstable substance. Therefore it is critical that you buy a form of SAM-e that has been processed and packaged correctly, and that you store it properly once you get it home.

How to Buy and Store SAM-e

Unlike over-the-counter and prescription medications, natural supplements and herbs are not regulated by the Food and Drug Administration to ensure their quality. Therefore, it is up to consumers to ask questions and to check the integrity and reputation of the supplement providers they choose to ensure they get a quality product. This is not always easy, because many people do not know what to ask. This situation exists for SAM-e, especially as it concerns which is the best chemical form of the supplement to take, because it is a highly unstable substance.

Let's explain what "stable" means. SAM-e is extremely hygroscopic, which means it absorbs moisture very easily. Once SAM-e comes in contact with moisture, it begins to degrade; thus it is referred to as "unstable." Because SAM-e is *so* hygroscopic, it is extremely unstable. What this means to you is that the form of SAM-e you buy must be manufactured and packaged properly to eliminate exposure to moisture, or its potency will be compromised, even to the extent that there may be no active SAM-e in the supplement at all. For this reason, it is recommended that you buy SAM-e that is enteric coated and packaged in blister packs.

The supplement SAM-e is available in two different chemical formulations: as butanedisulfonate and as tosylate (also referred to as sulfate). Based on detailed discussions with and information from experts who produce either form of SAM-e, as well as health professionals who have no vested interest in either form of SAM-e, the following is information that may help you when choosing which form of SAM-e to buy.

Both forms are highly unstable, and when tested by professional laboratories under highly controlled conditions for shelf life, both forms were similar: tosylate had a shelf life of thirty-six months and butanedisulfonate had a thirty-month life. For you as a consumer, the difference is basically unimportant, as it is unlikely you would store the supplement for any great length of time. Thus it appears that both forms of SAM-e are similarly stable when produced and packaged properly to protect against moisture. This information is supported by comments from Paul Packman, M.D., clinical associate professor of psychiatry at Washington University in St. Louis and past president of the American Academy of Clinical Psychiatrists. Noting that "both forms are hydrolyzed [broken down by water] into SAM-e," he says, "I can't think of a physiological reason as to whether the butanedisulfonate form or the tosylate form is better, because the bottom line is both end up as SAM-e in the body."

One difference between the butanedisulfonate and tosylate forms is their history. The butanedisulfonate form is a relative newcomer to the market, whereas the tosylate form is the one that has been used in the clinical trials over the last twenty-plus years. That is not to say that the butanedisulfonate form is not safe, but that the tosylate form has a longer record of safety on which to stand. Certainly, among physicians monitoring patients who are taking SAM-e, some have individuals taking the butanedisulfonate form, others the tosylate form. Positive reports are coming from people who are using both formulations, yet is still may be too early to tell if one form is truly superior to the other.

When it comes to which actual physical form of SAM-e to buy, there are some specific recommendations coming from experts in the supplement industry. Advocates of both forms of the supplement seem to agree on several points and differ somewhat on others. In that light, here are some things to consider when shopping for SAM-e:

- **Choose foil or foil blister packs.** Some makers of both the butanedisulfonate and tosylate forms say the tablets should ideally be packaged in foil (not plastic, which can let in moisture) or a foil blister pack. This protects them from the air and moisture, both of which will cause the tablets to oxidize (break down) and begin to lose their potency immediately. With a blister pack, you open one cell at a time as you need it.

- **Look for enteric coating.** An enteric-coated tablet has been coated with calcium carbonate, magnesium oxide, or another antacid, which prevents it from breaking down in the stomach and allows it to reach the intestinal tract, where it is dispersed into the bloodstream.

- **Choose tablets.** Tablets are reported to be superior to capsules. Bottled tablets, if enteric-coated and produced in a tightly controlled environment by a reputable manufacturer, are said to be nearly as potent as tablets in a blister pack. Loss of potency is not significant, only about 5 percent.

- **Don't be fooled by the price.** Do not assume that the more expensive product offers the maximum benefit. All SAM-e supplements are expensive; some are priced higher than others. Your selection should be based on the other items in this list and any information you get from physicians, pharmacists, and friends, because together they can determine the quality and your chances of getting results.

Until more comprehensive comparative studies are done on the two forms of SAM-e, it is suggested that you talk with your health-care provider, your pharmacist, and others who have taken SAM-e and follow the buying guidelines presented above. If you are like the majority of people who have tried SAM-e, you will get a response.

One very positive result of this controversy is that it is spurring on much research and development as producers try to create a more stable, more viable, and more economical SAM-e supplement. By mid-2000, a liquid form of SAM-e likely will be ready for the marketplace, and it reportedly is much more stable than the present solid forms. Thus advancements are being made every day, and hopefully the question as to which form of SAM-e is the best to take will be easier to answer very soon.

SAM-e Supplement Round-up

Brand	Enteric Coated	Form of SAM-e
General Nutrition Center	Yes	Butanedisulfonate
Great Earth Companies	No	Tosylate
Life Extension Foundation	Yes	Tosylate
Natrol	No	Tosylate
Nature Made	Yes	Butanedisulfonate
NutraLife Health Products	Yes	Tosylate
Solgar Vitamin & Herb	Yes	Tosylate

As of this writing, only Nature Made and NutraLife are available in blister packs. This is a representative list of suppliers; because of the popularity of SAM-e, new suppliers are often being added to this list. Available forms (tablet, capsule, blister pack) will also change.

Once you get SAM-e home, store it in a cool, dry place. A cabinet in the kitchen or bathroom is fine. Do not put it in the refrigerator, where there is a greater chance it will be affected by moisture. Only remove the tablets as you take them. Do not remove them from the blister pack and put them in your pocket or purse to take later. Take the entire blister pack with you if you are taking SAM-e away from home.

How Much SAM-e Should You Take?

When checking with physicians who are using SAM-e with their patients, it becomes clear that although there are guidelines for its use, dosages, especially when treating depression, can be highly individual. Because SAM-e is very safe (see information about safety below), individuals who have mild depression or osteoarthritis who take the recommended dose generally do not have any problems. Naturally, if you have any questions or concerns, always consult your physician.

Treating Depression

Here are two general recommendations for treating depression.

- Begin with 1,600 mg a day in divided doses: 800 mg twice a day or 400 mg four times a day. Take SAM-e on an empty stomach. Follow this program for two to three weeks, or until you begin to feel less depressed.
- Once you begin to feel better, gradually decrease your dosage to 800 mg or 400 mg daily.

Not all doctors agree with this dosage recommendation. Here is an alternative for those who may experience some gastrointestinal upset from SAM-e:

- Start at 200 mg twice daily on days one and two.
- Increase to 400 mg twice daily on days three through nine.
- Increase to 400 mg three times daily on days ten through nineteen.
- Increase to 400 mg four times daily after day twenty if needed.

It is important to remember that every person is unique, and that SAM-e's far-reaching effects in the brain can yield different results in different people. Your results will depend on the severity of your depression, if you are taking another antidepressant or other medications and the type you are taking, any other medical conditions you may have, and even your diet and the expectations you have about using SAM-e.

Gabriel Cousens, M.D., a licensed medical doctor, psychiatrist, and family therapist in Patagonia, Arizona, has used SAM-e for more than six months on more than forty patients with depression. He emphasizes the need to tailor the dose according to the needs of the individual. He generally starts patients on a dosage of three 200-mg tablets four times a day and then may increase it to levels higher than those recommended in this book. He has had a success rate in the 90 percent range for treatment of depression. He notes, however, that he uses it with other substances, as discussed below.

Taking SAM-e and Another Antidepressant

Some physicians find that giving SAM-e along with a conventional or natural antidepressant has several benefits (except for those patients with bipolar depression). If patients are taking a pharmaceutical antidepressant, it often allows them to gradually reduce their dependence on the drug. As their use of the drug declines, so does the occurrence of side effects from the medication. Whether SAM-e is taken along with a conventional antidepressant or a natural one (e.g., 5-HTP, St.-John's-wort), it can enhance the benefits of the other antidepressant.

Dr. Cousens notes that he uses SAM-e with other natural substances only, such as 5-HTP and tryptophan, depending on what the individual needs. "It [SAM-e] supports an overall program. The extra methylation somehow makes all the biochemistry of the brain work right," he says.

Here are some guidelines you can review with your physician for taking SAM-e along with another antidepressant. You may adjust this schedule to fit your individual needs.

- Begin with 1,600 mg per day of SAM-e.
- Gradually reduce the dosage to a maintenance level (400 to 800 mg daily) once you begin to experience the antidepressant effects. You may also be able to reduce the dosage of the other antidepressant or even eliminate its use altogether. Of course, always consult your physician before changing how you take any of your antidepressant medication.

More and more people are interested in following an all-natural (or as natural as possible) approach to the treatment of depression. SAM-e can be an important part of that treatment. Here are a few of the natural antidepressants that are most commonly used and their dosage suggestions. (See Chapter 4 for details about these natural antidepressants.) All of these supplements can be used safely and effectively with SAM-e; however, it is best to take them under a doctor's supervision.

- **St.-John's-wort.** This herb is available in many forms, but the most effective and easiest ones to take are the capsules, tincture, or extract. If using capsules, look for the standardized form at 0.3 percent hypericin (the active ingredient in St.-John's-wort). The recommended dosage is one 300-mg capsule three times daily, but you can start out more slowly by taking only two capsules daily and then increasing to three after a few days to a week. For the tincture, take four dropperfuls in water in the morning and three in the evening. If you do not notice significant results after two to three weeks, increase the dosage to one teaspoon in the morning and evening. St.-John's-wort is suggested for mild depression. More severe forms usually require different treatment.

- **5-HTP.** 5-HTP comes in tablets. The recommended dose for depression is 100 to 200 mg three times a day. 5-HTP requires adequate amounts of magnesium, niacin, and vitamin B_6 to convert to serotonin. The supplements recommended in the sidebar on page 186 are all excellent complements to this natural antidepressant. 5-HTP causes side effects in less than

10 percent of people who take the supplement, with nausea and dry mouth being the most common responses.

- **Ginkgo biloba.** Look for capsules, tablets, tincture, or extract standardized for 24 percent flavoglycosides and 6 percent terpene lactones (the active ingredients). Dosages range from 120 to 160 mg two to three times daily for tablets and capsules; 0.5 ml three times daily for the tincture; and 40 to 80 mg three times daily for the extract. Do not take ginkgo if you are pregnant or nursing or if you have a clotting disorder.

Treating Osteoarthritis

Again, not all doctors agree about the dosage schedule for SAM-e. Much depends on your individual needs. One recommended program for treatment of osteoarthritis is as follows:

- Begin with 800 mg daily in divided doses (400 mg twice a day) on an empty stomach. Take this dose for two weeks.
- Reduce the dosage to 400 mg daily, which should be taken once, first thing in the morning.
- If you tend to experience gastrointestinal distress from medications or if you are elderly, take the 400 mg in divided doses.

Here's another suggested dosage schedule.

- Start at 200 mg twice daily on days one and two.
- Increase to 400 mg twice daily on days three through nine.
- Increase to 400 mg three times daily on days ten through nineteen.

- Beginning on day twenty, reduce to 200 mg twice daily or as needed.

Other natural treatments for osteoarthritis include glucosamine and chondroitin, MSM, and stinging nettles. Suggested dosages of these supplements are explained below. They can be taken safely and effectively with SAM-e. (See Chapter 5 for more details on these supplements.)

- **Glucosamine and Chondroitin.** Glucosamine and chondroitin are available as tablets, capsules, and even as a flavored powder that can be mixed in water. If you are taking glucosamine alone, the recommended dosage is one 500-mg capsule or tablet three times daily, or a comparable amount of the powder (according to package directions). If you are taking both supplements, buy a combination product that contains 500 mg of glucosamine and 400 mg chondroitin in each capsule or tablet. The daily dosage of glucosamine and chondroitin should be in a 5:4 ratio; that is, 1,500 mg glucosamine and 1,200 mg chondroitin daily. This dosage should be increased proportionately for people who weigh more than two hundred pounds. Side effects associated with glucosamine are few and mild, generally stomach distress.
- **MSM.** MSM is available in capsules, powder, and tablets. For osteoarthritis, the suggested dosage is 2,000 to 6,000 mg daily in divided doses, depending on the severity of your symptoms. Take with food to avoid any possible stomach distress. If you do experience gastrointestinal discomfort, reduce your dosage for a few

days and then gradually build back up to your previous level. Occasionally MSM causes mild headache when people first start to take it. Drinking at least eight ounces of water with each dose of MSM can help relieve this side effect, which is frequently temporary.

- **Stinging nettles.** The suggested dosage is one to two 480-mg leaf capsules taken two to three times a day, or one to two 250-mg leaf extract capsules two to three times daily. Begin with the lowest dose and gradually increase until you get relief. Occasionally stinging nettles can cause stomach upset or a burning sensation on the skin.

Treating Other Conditions with SAM-e and Other Natural Remedies

Suggested dosage programs for treating other conditions with SAM-e are below. Consult with your health-care provider to determine the best treatment plan for you. If you have Parkinson's or Alzheimer's disease or any other medical condition, it is especially important that you only take SAM-e under the supervision of your health-care provider.

- **Fibromyalgia:** 200 to 400 mg twice daily. In Chapter 6 a natural remedy combination for fibromyalgia was discussed. Below is an explanation of how to use that treatment. It can be taken after meals. Consult first with your physician when combining this remedy with SAM-e.
  - **St.-John's-wort.** Take the 300-mg capsules that have been standardized at 0.3 percent hypericin. The recommended dosage is one

capsule three times daily. Take the capsule along with 5-HTP and magnesium.

- **5-HTP**. Take one 50- or 100-mg tablet three times a day along with St.-John's-wort and magnesium. Whenever taking 5-HTP, be sure you are getting enough niacin and vitamin B_6, which are easiest to get if you are taking a multivitamin-mineral or vitamin B complex supplement.
- **Magnesium**. Take one 150- to 250-mg tablet three times a day along with St.-John's-wort and 5-HTP. Use the magnesium citrate, malate, fumarate, or succinate forms.
- **Liver disease and migraine**. Although there are several effective natural remedies for both migraine and liver disorders, including feverfew, ginger, and 5-HTP for migraine and milk thistle for liver conditions, there are no published studies of the use of SAM-e with any of these supplements, so none are suggested here. The recommended dosage of SAM-e for migraine and liver disease is given below.
  - **Liver disease:** 200 to 400 mg two to three times daily on an empty stomach.
  - **Migraine:** 200 to 400 mg twice daily on an empty stomach.

Supplements to Take with SAM-e

Regardless of how much SAM-e you take or for what reason, you should always make sure you are getting enough folic acid, vitamin B_6, and vitamin B_{12} either in your diet or as a supplement. These nutrients are essential for many reasons, but they are especially important when taking SAM-e. Here are a few of the reasons why:

- Most importantly, SAM-e needs these nutrients to perform its vital functions. For example: folic acid helps boost the methylation cycle, and vitamin B_{12} helps in the production of methionine. Both folic acid and vitamin B_{12}, like SAM-e, are methyl donors, which means they donate to neurotransmitters like dopamine and serotonin and help relieve depression.

- Low levels of folic acid, like low levels of SAM-e, are associated with a higher risk of depression (see Chapter 3). Depression is the most common symptom of a folic acid deficiency.

- Although vitamin B_{12} deficiency is less common than folic acid deficiency, it is an especially important cause of depression among the elderly. Other signs of vitamin B_{12} deficiency in older people can show up as problems with balance, memory loss, dementia, and tingling in the hands and feet.

- Symptoms of a deficiency of vitamin B_{12} (e.g., anemia, nervousness, heart palpitations, mouth sores) can be masked by folic acid. Therefore, you should always take a vitamin B_{12} supplement when taking folic acid.

- Vitamin B_6 levels are typically low in people who are depressed and in women who take any form of estrogen therapy, including birth control pills. Because B_6 is needed to produce serotonin, many depressed people may need to take vitamin B_6 supplements to improve their depressive symptoms. Other symptoms associated with a deficiency of vitamin B$_6$ are irritability and sensitivity to sound.

Because these three nutrients are often deficient in people's diets, it is best to take a supplement, either separately or in a multivitamin-mineral. When taking SAM-e, it is recommended that you take the following dosage of these three nutrients: 800 mcg folic acid, 800 to 1,000 mcg vitamin B$_{12}$, and 100 mg vitamin B$_6$. You may find all three in sufficient amounts in a vitamin B complex supplement or a multivitamin-mineral.

Optimize SAM-e Effectiveness

For optimal results when treating depression with SAM-e, you can take a B-complex supplement along with the other vitamins and minerals listed in the sidebar. Collectively, these nutrients can help fight depression, stress, and anxiety and provide a tremendous boost to your SAM-e supplement. Although not every B vitamin is especially beneficial for depression (you will notice that several are missing in the sidebar listing), it is best to take all B vitamins together because they work as a unit.

How Safe Is SAM-e?

Drugs and side effects: it's an unfortunate fact of life that the two go together. Sometimes the side effects caused by a drug are so bad that people decide to quit the drug rather than suffer with its consequences. These are cases in which the treatment is worse than the cure.

One of the best advantages of taking SAM-e is that unlike most prescription and over-the-counter drugs,

Nutrients that Fight Against Depression, Stress, and Anxiety

Nutrient and Dosage	Action
Vitamin B₁ (thiamine) 100 mg	Helps alleviate depression and anxiety
Vitamin B₆ (pyridoxine) 100 mg	Aids in the manufacture of the neurotransmitters dopamine and norepinephrine
Vitamin B₅ (pantothenic acid) 100 mg	Relieves tension
Folic acid 800–1,000 mcg	Low levels are associated with depression
Vitamin B₃ (niacin)*	Essential for proper functioning of the nervous system
Choline 100 mg	Helps send nerve impulses to the brain
Vitamin B₁₂ 1,000 mcg	Relieves irritability, improves concentration
Vitamin C 1,000–3,000 mg	Fights against stress
Vitamin E 400 IU	Helps get oxygen to the brain cells
Calcium 1,000 mg	Relieves tension and irritability
Magnesium 500 mg	Necessary for proper nerve functioning
Manganese 15 mg	Helps reduce nervous irritability
Zinc 15 to 30 mg	Promotes mental alertness

*To avoid the flush that often accompanies niacin supplementation, take the niacinamide form, 30 mg daily.

it causes minimal or no side effects. In fact, in placebo-controlled trials, people taking the placebo often
experienced more adverse reactions than the people
taking SAM-e. Many physicians and researchers hail
this lack of significant side effects as one of the most,
if not *the* most, appealing thing about SAM-e.

To test the toxicity of a substance (its ability to
cause harm to organs or to systems in the body), researchers usually administer extremely high doses of
it to laboratory animals. In the case of SAM-e, studies
reported in 1987 found that rats given doses as high
as fifty times the maximum dose given to humans
(which is 1,600 mg daily) did not exhibit any notice-
able damage to their chromosomes. In another study,
pregnant rats were given doses more than ten times
the maximum human dose, and both the mothers and
fetuses were seemingly unharmed. A similar study
done in rabbits had the same results.

But rats and rabbits are not people, and the only
true test of the toxicity of a substance in humans is to
give it to them. In Europe, physicians and patients
have enjoyed the fact that SAM-e does not cause the
distressful adverse reactions associated with conventional antidepressants (e.g., sexual problems, anxiety,
weight gain, nausea, rash) or those caused by medications for osteoarthritis (bleeding of the stomach
lining, ulcers). That said, occasionally people do
experience a mild headache when they first start taking SAM-e, but it generally only lasts a few days. At
high dosages, SAM-e can cause mild stomach distress
or heartburn. If you experience these stomach problems, you can reduce your dosage for a few days and
then gradually bring it back up or drink more water
when taking the tablet (eight ounces of water is sug-

gested). If these measures fail, try taking SAM-e with food.

Dr. Sol Grazi, in his book *The European Arthritis and Depression Breakthrough! SAMe* says that "SAMe should be only one part of an overall approach that involves self-help strategies and various other therapies as appropriate." This is sound advice. Perhaps the phrase "other therapies" should be expanded to include wise lifestyle choices, such as a healthy diet, regular exercise, adequate sleep, stress management, no smoking, little or no alcohol, social interaction, and avoidance of drugs. The body is a holistic system comprised of physical, emotional, and spiritual components. A failure to attend to the needs of all three components results in an unbalanced state of being, which leads to illness. When applied in the correct circumstances, SAM-e can address physical or emotional needs, or both. It can complement other therapies or replace them, depending on the situation.

If you decide to give SAM-e a chance, make sure you give it your best shot. That means purchasing an effective form, taking the appropriate nutritional supplements, and consulting a physician to help you determine the treatment plan that best applies to your situation. If you are battling depression, that may mean combining counseling or participation in a self-help group, or taking another antidepressant along with SAM-e. If osteoarthritis is keeping you down, follow your doctor's advice about proper diet and regular exercise. If liver disease is your concern, make every effort to eliminate toxins from your diet and your lifestyle, including alcohol, cigarettes, pesticides, and food additives. When you add SAM-e to your treatment plan, it may be the partnership you've been waiting for.

NOTES

Carney, W.M.W.P., Williams, D.G., and Sheffield, B.F. "Thiamin and pyridoxine lack in newly-admitted psychiatric patients." *British Journal of Psychiatry* 135 (1979): 249-254.

Cousens, Gabriel. Telephone interview with author, 21 June 1999.

Cowley, G., and Underwood, A. "What is SAMe?" *Newsweek*, 5 July 1999, pp. 26–50.

Cozens, D.D., et al. "Reproductive toxicity studies of ademetionine." *Arzneitmittel-Forschung* 38 (110: (November 1988): 1625–1629.

Kivela, S.L., Pahkala, K., and Eromnen, A. "Depression in the aged: Relation to folate and vitamins C and B_{12}." *Biol Psychiatry* 26 (1989): 209–213.

Mindell, Earl. *Earl Mindell's Vitamin Bible.* New York: Warner Books, 1991.

Murray, Michael. *Encyclopedia of Natural Medicine.* Rocklin, CA: Prima Publishing, 1998.

Pezzoli, C., Galli-Kienle, M., and Stramentinoli, G. "Lack of mutagenic activity of ademetionine in vitro and in vivo." *Arzneitmittel-Forschung* 37 (7) (July 1987): 826–829.

Reynolds E., et al. "Folate deficiency in depressive illness." *British Journal of Psychiatry* 117 (1970): 287–292.

Reynolds, E., and Stramentinoli, G. "Folic acid, S-adenosylmethionine and affective disorder." *Psychol Med* 13 (1983): 705–710.

Telephone interviews: David Dillon, President, NutraPure; Marshall Fong, Customer Service, Pharmavite; Jonathan Glass, Director, Product Development, NutraLife Health Products; Paul Packman, M.D., Clinical Associate Professor of Psychiatry, Washington University in St. Louis.

Zucker, D., et al. "B_{12} deficiency and psychiatric disorders: A case report and literature review." *Biol Psychiatry* 16 (1981): 197–205.

10

Questions about SAM-e

This chapter includes questions about SAM-e that were not answered elsewhere in this book, as well as those that were addressed in the preceding chapters but which are easier to find the answers to here. (For questions of the latter type, you are referred to the appropriate chapter for more details.)

Why should I take SAM-e for depression?

The main benefit of taking SAM-e for depression appears to be its ability to enhance the levels and metabolism of the neurotransmitters serotonin, dopamine, and norepinephrine. SAM-e also has the ability to work more quickly and without the side effects of prescription antidepressants. See Chapters 3 and 4 for details about SAM-e in the treatment of depression.

Why should I take SAM-e for osteoarthritis?

The best reason to take SAM-e for osteoarthritis is that it appears to stop and reverse the degeneration of the cartilage in the joints—to our knowledge, no prescription drug on the market has this ability. An added bonus is that it does not cause the gastrointestinal problems associated with the use of nonsteroidal anti-inflammatory drugs. See Chapter 5 for details about SAM-e and osteoarthritis.

What's the number one reason to take SAM-e for liver disease?

SAM-e is a precursor for the production of the antioxidant glutathione. This potent substance is the primary protector of cells from free radical damage and the main component in the detoxification process in the liver of drugs, alcohol, environmental poisons, and other toxins. Without the ability to make glutathione or enough of it, as happens in people with cirrhosis, the liver becomes more and more damaged, the level of toxins in the body rises, and overall health deteriorates and can result in death.

Will SAM-e work for me?

If you have a condition that has been successfully treated with SAM-e—depression, osteoarthritis, fibromyalgia, or liver disease—chances are you will get some benefit from taking SAM-e. If you have Alzheimer's disease or Parkinson's disease or another condition discussed in this book, the evidence is less solid. In any case, consult with your doctor before taking the supplement. But do not take SAM-e simply because you *think* you have low levels of the substance in your body. You will probably be wasting your time and money.

Why can't I get all the SAM-e I need from my diet?

Normally, the body produces all the SAM-e it needs from methionine, an amino acid that is found in many protein-rich foods such as soybeans, seeds, lentils and other legumes, eggs, and meat. But there are many situations that can cause SAM-e levels to be low, including being elderly or having depression, osteoarthritis, Alzheimer's disease, liver disorders, fibromyalgia, and other medical conditions. In addition, a diet that is low in vitamin B_{12}, folic acid, choline, or methionine can impair the body's ability to produce SAM-e

Why can't I take supplements of the amino acid methionine alone instead of taking SAM-e?

Although this sounds like a good idea, studies have shown that taking high doses of methionine does not increase the body's levels of SAM-e, nor does it provide the benefits of SAM-e. In fact, high doses of methionine can be toxic. Generally, it is not recommended that you ever take supplements of single amino acids at high doses for any extended period of time. The balance of all the amino acids in the body needs to be maintained at a certain level, and when you take a supplement of just one or two amino acids, you throw off that balance. Occasionally people do take individual amino acid supplements but should only do so under the supervision of a health-care provider.

I noticed that many of the early studies used an injectable form of SAM-e. What's the difference between oral and injectable SAM-e? Are they equally effective?

Nearly all of the early studies of SAM-e used intramuscular, subcutaneous, or intravenous injections of SAM-e. When an oral form became available, scientists made adjustments in the dosages to account for the difference in how injectable versus oral forms are assimilated by the body. The two forms are equally effective, and the oral form is obviously much preferred by and more convenient for consumers.

To determine the optimal way to take SAM-e supplements, researchers tested the absorption rate and efficiency of oral forms. Using animal models, they found that a dose of SAM-e administered directly into the intestines (intraduodenally) resulted in a six times higher concentration of the substance than did an oral dose. Naturally, it is not practical for you to take SAM-e intraduodenally. So researchers tested the efficacy of SAM-e in an enteric-coated capsule in dogs. An enteric coating is a substance that prevents the contents of the capsule or tablet from being broken down by the digestive acids in the stomach and thus allows the supplement to reach the intestines, where it can be dispersed to the bloodstream.

What should I do if I miss taking a dose of SAM-e? Should I take twice as much at my next dose?

If you miss your morning dose, take it before lunch. If you forget both doses, do not double up the next day, just take your normal dose.

Why do I have to take all of my SAM-e doses in the morning? Can't I take a dose at night before I go to bed?

Some people report feeling a surge of energy after taking SAM-e. If you take a dose at night, you may have trouble falling asleep. If you want to change your dosing schedule, you can take your second dose in the afternoon if you feel you need a boost of energy.

I'm used to taking vitamins after I eat, so taking SAM-e on an empty stomach is hard to remember. Do you have any tips?

Some people like to take it as soon as they get up in the morning: it's next to their toothbrush and they take it even before they take a shower. If this doesn't fit your morning routine, perhaps if you put it on your nightstand you'll see it first thing in the morning and take it then. For your before-lunch dose, take it when you start to get hungry, or try to schedule your dose with something you normally do about thirty minutes or so before lunch each day. You can also write yourself little notes and stick them where you can't miss them, like on your computer screen or the dashboard of your car.

What happens if I accidentally take too much SAM-e?

No one has ever reported a death associated with taking an overdose of SAM-e. Some people have taken up to 3,600 mg a day without experiencing any adverse effects. Stomach and intestinal distress can occur at the higher end of the recommended dosage (1,600 mg). However, there is no reason to take more than the recommended dosage for the condition you wish to treat (see Chapter 9). Studies show that benefits from SAM-e are obtained at dosages of 1,600 mg or less per day.

Is it safe to drink alcohol while I'm taking SAM-e?

Generally, there is no reason why you can't drink alcohol while taking SAM-e. However, if you are treating a liver disorder, alcohol should be avoided. If you are treating depression, alcohol consumption is not recommended because it is a depressant.

Is SAM-e safe to take if I'm pregnant or nursing?

No specific studies have been targeted to that question, so the safe answer is no. Here's what we know so far. Pregnant women have taken SAM-e (up to 1,600 mg daily) for short periods of time and have not experienced any problems for themselves or their infants. Also, infants naturally have very high levels of SAM-e, so it is assumed that whatever amount they may get through the breast milk is safe. However, you should consult your physician before taking any supplement or drug while you are nursing, and that includes SAM-e.

One important note of caution: if you want to take SAM-e because you are experiencing postpartum depression, get a definitive diagnosis from your physician before you decide to self-medicate. Up to 15 percent of women with postpartum blues have bipolar depression, and SAM-e is not recommended as a treatment for that type of depression. (See Chapter 3 for more on postpartum depression.)

Is SAM-e safe to give to children?

SAM-e has been given to children in only a few studies and for very specific conditions. Most parents ask about SAM-e as a treatment for depression in their children. Although there are no indications from other studies that SAM-e would be harmful to children, you should consult with your child's physician, get an accurate diagnosis, and discuss whether it is advisable to treat the child with SAM-e.

What side effects does SAM-e cause?

Most people don't report any adverse reactions. Some people experience stomach distress; others get a mild headache or heartburn. (See Chapter 9 for more details on side effects and toxicity.)

What is the difference between toxicity and side effects?

Toxicity is the ability of a substance to damage organs or body systems; that is, how poisonous a drug or other substance is to the body. Side effects are the reactions people have to the poison; for example, headache, nausea, dry mouth, or vomiting. The more toxic a substance is, the more severe the side effects will be.

I saw SAM-e referred to as AdoMet. Is that something different?

No. SAM-e's "real" name is S-adenosylmethionine, also written as S-adenosyl-L-methionine. You will also see SAM-e written as SAMe, SAM, AdoMet, ademe-

tionine, and even Sammy. They are all the same substance.

Where can I buy SAM-e?

As SAM-e gains in popularity, it is available in more
and more places, including pharmacies, grocery
stores, discount stores, major chain stores, through
mail order, and over the Internet. No matter where
you buy SAM-e, however, remember that it will not
be effective if it has not been processed or packaged
correctly. Only buy from reputable manufacturers.
(See Chapter 9 on how to buy and store SAM-e.)

Why is SAM-e so expensive?

SAM-e is a highly unstable substance, which means it
oxidizes (breaks down) rapidly when exposed to air
and moisture. Therefore the manufacturing process is
very specialized and must be highly regulated to ensure a potent product. That process is not cheap. In
addition, it is necessary to individually package the
supplement to ensure its potency. This adds to the
cost as well. It is hoped that as SAM-e becomes more
available in the United States the price will come
down.

What is the normal amount of SAM-e the body should have?

As Dr. Grazi explains in his book *The European Arthritis and Depression Breakthrough! SAMe*, there is no
direct relationship between having a low level of
SAM-e and the amount of benefit you will get from

taking SAM-e supplements. When you take SAM-e, you aren't so much increasing the level of SAM-e in the body as you are increasing the number of methyl donors that are available to help in the dozens of processes in which SAM-e is involved (see Chapter 2). The body is inherently "smart" and knows to distribute SAM-e where it is needed most.

How can I find a doctor who is familiar with SAM-e?

Because SAM-e is so new to the United States, you may not find many physicians who are well versed in SAM-e and how it works. What can be just as important, however, is a physician who is open-minded and willing to learn about how SAM-e can help his or her patients. If you have such a physician, you can discuss this book and other information about SAM-e with him or her. More studies of SAM-e are coming out all the time, and so you will likely have additional information you both can share.

If your health-care provider is not willing to talk about SAM-e with you, then you need to look for another physician. Physicians who may be more open to using SAM-e or who may be recommending it to their patients are those who specialize in or already use other complementary therapies in their practice, physicians who are also homeopaths, and physicians who are associated with the American College for the Advancement of Medicine. See the Appendix for information on how to contact physicians in these categories.

If raising low levels of SAM-e helps depression, osteoarthritis, and liver disease, doesn't it make sense that

it should help any condition in which levels of SAM-e are low?

On the surface, that concept sounds logical, and it would be wonderful if it were true. You need to consider two things, however. One, SAM-e is involved in many very complex processes, all of which involve other components in the body. Each of these components depends in some way on many others to perform its designated tasks. And two, our bodies are extremely complicated organisms, full of feverish biochemical processes and mysterious mechanisms. SAM-e participates in at least forty processes, but these are only a small part of what is happening every second in the body. Just think, methylation occurs a billion times a second, every second of your life!

It is fortunate that scientists have found a positive correlation between SAM-e and depression, osteoarthritis, liver disease, and fibromyalgia. In each of these conditions, supplementing with SAM-e provides different benefits, and there may be others that researchers have not even discovered yet. It is also promising that there is good evidence SAM-e may be helpful in the treatment of other medical conditions. But it does not mean that low levels of SAM-e are always necessarily bad and that doctors should rush in and "fix it." See the discussion on Parkinson's disease in Chapter 8 as an example.

When you raise the level of any substance in the body, you ultimately trigger any number of events and responses—some you expect, others you may not. As you've learned in this book, SAM-e is involved in some very intricate processes in the body. In the case of SAM-e and depression, osteoarthritis, fibromyalgia, and liver disease, the triggered responses have been beneficial for the most part. Many people feel better and get better. But there are those people who don't respond at all, or perhaps only minimally. Why? If two people of the same age, with similar physical characteristics and lifestyles, both have osteoarthritis, for example, why will one person get tremendous relief from a treatment such as SAM-e (or glucosamine and chondroitin, for that matter) and the other get little or no relief?

The answer may lie somewhere in the complex network of nerves and tissues and organs and cells. It may be lurking deep in the recesses of the brain. Part of the answer could be that one person has a genetic makeup that makes him or her more prone to osteoarthritis or that one has an undetected food allergy to nightshade vegetables, which have alkaloids that can inhibit normal collagen repair. It's the same story with any drug, any herb, any nutritional supplement. Not everyone responds in the same way. There are expected responses, but there are always exceptions to the rules.

The truth is, it's a gamble. Every treatment, every remedy, every therapeutic method is a toss of the dice. The job of researchers and physicians is to improve the odds in favor of the house—your house and the houses of the majority of people who have a specific condition. The scientists investigate and evaluate and quantitate to find the most benefit for the most people while leaving as little as possible to chance. Thus far SAM-e has proven to be a winner in several "games," and as research continues it is hoped it will be a winner in many more. If it is not, then we still have a winner.

Why did SAM-e first become so widely used in Europe instead of in the United States?

There are several parts to the answer to this question. One concerns attitude and tradition. The institution of Western medicine in the United States is strongly planted in the conventional, scientific method and has largely eschewed anything outside that realm. The Europeans, however, have remained open to both modern medical practices and traditional, natural remedies. This attitude on the part of the Europeans supports and nurtures the work of their scientists and physicians to conduct research into herbal and homeopathic remedies, nutritional therapies, and other natural healing practices. They are much more willing to integrate the Western medical system with other medical models. It is largely their dedicated research that has brought us St.-John's-wort, ginkgo biloba, echinacea, feverfew, and many other natural remedies. And now SAM-e.

The second part of the answer concerns economics. The United States has laws, formed and regulated by the Food and Drug Administration (FDA), that govern how, when, and under what conditions drugs can be made available to the marketplace. These laws include the need for pharmaceutical companies to invest many years and much money in testing any drug they hope to put on the market. Under these laws, it is also very difficult for pharmaceutical companies to get a patent on a natural substance (e.g., SAM-e, or any herbal remedy). Because pharmaceutical companies are in business to make a profit, and there is virtually no incentive for them to gamble time and money on a natural product, such supplements have been largely ignored by U.S. drug companies.

Fortunately, the attitude of those who practice Western medicine in the United States is changing, and it is changing because consumers like yourself are demanding natural remedies. People have found that Western medicine often does not have the answers they need and that many of the natural remedies work just as well or better than their drug counterparts. Consumers are also tired of suffering with the side effects of many pharmaceuticals that are often more distressful than the conditions they are designed to treat.

The attitude among the Western medicine establishment is also changing. In 1994, Congress passed the Dietary Supplement Health and Education Act. This act allows the FDA to approve a nutritional supplement as long as there is reasonable evidence that the supplement is not harmful. That same act is the reason why SAM-e can now come to your pharmacy and grocery store shelves.

Can SAM-e help me lose weight?

To date, there do not appear to be any studies on SAM-e and weight loss. However, if you're depressed and don't feel like exercising and then take SAM-e and your mood improves, you may be motivated to start an exercise routine. Or, if you tended to overeat when depressed and your mood improves, you may eat less and lose weight. However, no direct link has been made between taking SAM-e and weight loss.

Can I take SAM-e with other supplements?

SAM-e appears to be safe to take with other supplements, including vitamins, minerals, coenzymes, amino acids, hormones, and herbs. In fact, several of the B vitamins, including folic acid and vitamins B_6 and B_{12}, need to be taken along with SAM-e (see Chapter 9). If you want to add SAM-e to a treatment program that includes St.-John's-wort, glucosamine, or other natural supplements, see Chapter 9. It is recommended that you consult with your health-care practitioner for the best dosage schedule and combinations that best suit your needs when taking SAM-e and other supplements.

Can I take SAM-e while I'm also taking prescription medications?

Thus far, researchers have not found any reason you cannot take SAM-e with conventional medications, except in the case of monoamine inhibitors. In fact, SAM-e can be taken to enhance the benefits of other antidepressants (see Chapter 4).

My doctor says SAM-e works by "loading the system." What does that mean?

Many substances are made or used naturally by the body; for example, SAM-e, glucosamine, and melatonin. Some of these substances can be taken as supplements to enhance the processes in which they are involved. When you take these supplements, you are "loading" your system with raw material it can use to participate in the activities it normally does. When you take SAM-e, for example, you enhance the amount of methyl groups that are donated, which increases various methylation processes, processes that can result in improved cell permeability, higher neurotransmitter levels, lower homocysteine levels, restoration of cartilage, and dozens of other beneficial responses. Thus, when you load your system, you are giving it the tools it needs to make the adjustments your body requires.

Epilogue

When SAM-e hit the U.S. market in February 1999, the general public had a barrage of questions about the new supplement and the claims being made about it. The National Arthritis Foundation was inundated with so many inquiries about SAM-e and its ability to relieve pain that the association finally released a statement saying that there was sufficient evidence that SAM-e did provide pain relief. At the same time, the foundation cautioned that people should not take this as carte blanche to quit their regular treatment programs and that there was insufficient proof that SAM-e restored cartilage. These are reasonable and expected precautions to come from a conventional medical establishment. To give a blanket endorsement of SAM-e at this point would be completely irresponsible of any organization, or individual for that matter. Right now, we are in a "wait and see" stage as investigations continue and more and more physicians introduce it to their patients.

A *Los Angeles Times* article on 14 June 1999 on the phenomenon of SAM-e was entitled "Magic Pill or Minor Hope?" It is the hope of many people who suffer with depression, osteoarthritis, liver disease, fibromyalgia, Alzheimer's disease, Parkinson's disease, and a variety of other disorders that the proof sways more toward the former than the latter.

The SAM-e Solution has brought you the latest information available about SAM-e at the time the book went to press. It has shown you that SAM-e is not a silver bullet, nor is it the answer to all of your medical problems. Nor, for that matter, are penicillin, aspirin, Prozac, Valium, or St.-John's-wort. Some solutions work better than others; SAM-e appears to be one of those better solutions. Millions of Europeans during the last twenty-five-plus years have made it part of their solution. Now it may be time for Americans to make it part of theirs.

Glossary

Acetaldehyde: A byproduct of the metabolism of alcohol.

Acetaminophen: An over-the-counter pain reliever that does not contain aspirin. It helps relieve pain but not inflammation

Acetylcholine: A chemical that transmits impulses between nerves and between nerves and muscle cells.

Adrenal glands: Glands, located near the kidneys, that synthesize and store dopamine, norepinephrine, and epinephrine.

Aneurysm: A bulging or dilatation of the arteries or veins.

Antidepressants: Drugs used to relieve depression.

Atheroma: A thickening or deterioration of the walls of the arteries in atherosclerosis.

Atherosclerosis: A condition in which there is a buildup of plaque in the blood vessels that can cause heart attack.

ATP (adenosine triphosphate): A molecule that is used in the cells to store energy. It combines with methionine to make SAM-e in the body.

Autoimmune disease: A disease in which the body's immune system perceives normal parts of the body as foreign and attacks them. Examples of autoimmune conditions include allergies and rheumatoid arthritis.

Bilirubin: A byproduct of the hemoglobin molecule of red blood cells. It is carried to the liver in the bloodstream and converted in the liver, then excreted in the bile.

Bioavailability: The degree to which a substance is available for use by the body.

Carbidopa: A drug used along with levodopa to treat Parkinson's disease.

Cartilage: A type of tough connective tissue that is part of joint structure. It also is found in the ears and nose.

Cholestasis: A condition in which the bile becomes stagnant in the liver.

Choline: A substance that is essential in the metabolism of fat and carbohydrates. It is also a precursor of acetylcholine.

Chondrocytes: Cells that make up cartilage. They manufacture collagen and proteoglycans.

Chondroitin sulfate: A substance found in the connective tissue and cell membranes. It attracts water to cartilage and helps maintain the elasticity and integrity of the body's tissues.

Chrondroprotective: A chondroprotective substance protects the chondrocytes and allows them to synthesize proteoglycan and collagen—and thus prevent damage to the cartilage.

Chronic fatigue syndrome: A condition characterized by long-term fatigue with no identifiable cause.

Cirrhosis: An irreversible liver disease in which the tissue is damaged by alcohol, viruses, or other toxins.

Collagen: A protein that is involved in the development of connective tissue.

Congestive heart failure: A chronic disease in which

the heart is not able to supply the oxygen the body needs.

Connective tissue: A type of tissue that provides the structure and support to the body.

Controlled study: A study in which one group of people or subjects is used as a standard for comparison against another group. The subjects themselves are known as *controls*.

Coronary artery disease: A condition in which the heart does not receive enough blood and oxygen because the blood vessels that supply them are blocked.

Corticosteroids: Hormones produced by the body that have anti-inflammatory effects.

Cortisol: A hormone involved in the metabolism of fats, carbohydrates, and proteins, and which also has anti-inflammatory properties.

Cortisone: A hormone produced by the adrenal gland that has anti-inflammatory effects.

Cysteine: A sulfur-containing amino acid.

Cytochrome P450 system: A group of more than one hundred enzymes that are involved in the neutralization of various chemical substances in the liver during Phase One detoxification.

Delta sleep: A stage of deep, restorative sleep in which the brain produces a wave called the delta wave.

Dementia: A loss of mental functioning.

Dopamine: A neurotransmitter that is made in the adrenal gland. It plays a primary role in Parkinson's disease and depression.

Double-blind study: A method of research in which two drugs or other types of intervention are compared and neither the subjects nor the people evaluating the findings know the identity of the intervention. This is done so none of the participants will be influenced by their opinions or expectations.

Enteric-coated: Tablets or capsules coated to ensure they do not dissolve in the stomach before they reach the intestinal tract.

Fatigue: A condition in which you feel as though you have no energy, or a general feeling of being worn down.

Fibromyalgia: A rheumatic condition that affects the body's muscles and other soft tissues. It is characterized by fatigue, muscle pain, and poor sleep patterns.

Framingham Heart Study: A landmark study, begun in 1947 and still continuing today, in which more than five thousand adults chosen from the general population were evaluated and followed over the years to determine risks for heart disease. Because extensive medical data have been kept on these individuals, they are often used as the basis for other studies in which researchers study risk factors and medical phenomena.

Free radicals: Molecules that have unpaired electrons and thus attach themselves to other molecules, take their electrons, and cause damage (known as *oxidation*) to the cells in the process. Free radicals are neutralized by antioxidants.

Glucosamine: A compound that is found naturally in the body and which helps hold tissue cells together and is part of the synovial fluid that cushions the joints.

HAM-D: The Hamilton Depression test is an objective measure of the severity of depression.

Homocysteine: A sulfur-containing amino acid that is a byproduct of the conversion of methionine.

Intrahepatic: Something that occurs or concerns the inside of the liver.

Irritable bowel syndrome: A chronic condition characterized by abdominal pain, constipation, and diarrhea, with the latter two symptoms often alternating in occurrence.

Jaundice: A condition in which the skin and whites of the eyes turn yellow because of an accumulation of bilirubin in the body.

Levodopa: Often abbreviated as L-dopa, it is a drug used to treat Parkinson's disease.

Lipids: A classification of substances that includes fats, steroids, prostaglandins, and phospholipids.

Lipotropic: Something that promotes the flow of lipids to and from the liver and thus prevents the buildup of fat in the liver.

Lupus: Also known as systemic lupus erythematosus, it is an inflammatory autoimmune disease that affects the connective tissues and can involve the skin, joints, blood, kidneys, and other organs.

MAO inhibitor: A drug that blocks monoamine oxidase, an enzyme that causes the breakdown of the neurotransmitters serotonin, dopamine, and norepinephrine.

Melatonin: A hormone produced by the pineal gland at night and which depends upon SAM-e for its production. It is instrumental in the sleep-wake cycle.

Meta-analysis: A study design that combines similar factors from smaller studies into one study in order to provide a broader perspective than individual studies can offer.

Methionine: A sulfur-containing essential amino acid that is also one of the two components that make up SAM-e.

Methyl donor: Any substance that gives up a methyl group (one carbon atom and three hydrogen atoms) in a chemical reaction called methylation.

Methylation: A process in which methyl groups are transferred from one molecule or compound to another.

Neurotransmitter: A chemical that is released from a nerve cell and sent to influence another nerve cell. Dopamine, serotonin, and norepinephrine are examples of neurotransmitters.

NSAID: The acronym for nonsteroidal anti-inflammatory drugs, which are used to relieve pain by reducing inflammation. They are available either over-the-counter or by prescription.

Osteoarthritis: A disease in which the cartilage breaks down in certain joints, causing pain, limited mobility, and deformity.

Plaque: In the arteries, a deposit of fatty substances.

Precursor: A substance that precedes another substance.

Prostaglandins: Hormone-like substances, manufactured from essential fatty acids, which can cause inflammation.

Rheumatic disease: A general term that describes conditions characterized by pain and stiffness of the muscles or joints. Although this term is sometimes used interchangeably with "arthritis," some rheumatic conditions do not involve inflammation, as do arthritic diseases.

Serotonin: A hormone that is involved in balancing mood as well as many other functions.

SSRI (selective serotonin reuptake inhibitors): A group of antidepressants that decrease the levels of serotonin in the brain. Prozac and Zoloft are two commonly used SSRIs.

Tender points: Specific sites on the body that are abnormally sensitive and that are painful when pressed. The presence of tender points is used to help diagnose fibromyalgia.

Varices: Twisted, dilated veins.

Organizations of Interest and Sources of SAM-e and Other Supplements

Sources of SAM-e

Look for SAM-e in pharmacies, nutrition and health food stores, grocery stores, large chain buying centers and department stores, and other retail outlets, as well as on the Internet and through mail order. Refer to the buying guidelines in Chapter 9 before making your purchase. The following is a partial list of distributors of SAM-e.

General Nutrition Centers
Check the telephone book for the store nearest you

Great Earth Companies
Sold through retailers
1-800-57-VITAMIN
http://www.greatearth.com

Life Extension Foundation
PO Box 229120
Hollywood, FL 33022-9120
1-800-841-5433
http://www.lef.org

Natrol
Sold through retailers
1-800-326-1520 (customer service)
http://www.natrol.com

Nature Made
Sold through retailers and over the Internet
Nature Made Healthline (to speak with a nutrition expert)
1-800-276-2878
http://www.naturemade.com/inside/index.html

NutraLife Health Products
1328 River Avenue, Suite 151
Lakewood, NJ 08701
http://www.nutralife.com
1-877-688-7254

Solgar Vitamin & Herb Company
Sold through retailers
http://www.solgar.com

For Information on Herbs and Supplements

American Botanical Council
PO Box 201660
Austin, TX 78720-1660
1-512-331-1924

Center for Science in the Public Interest
1875 Connecticut Avenue W, Suite 300
Washington DC 20009
1-202-332-9110

Food and Drug Administration
5600 Fishers Lane, HFE 88
Rockville, MD 20857
1-301-443-3170
(or see the Yellow Pages for your regional FDA office)
http://www.fda.gov

The Herb Research Foundation
1007 Pearl Street, #200
Boulder, CO 80302
1-303-449-2265

For more information about the medical conditions discussed in this book, refer to the following organizations and/or websites.

Alzheimer's Association
919 N. Michigan Avenue, Suite 1000
Chicago, IL 60611
1-800-272-3900
http://www.alz.org

Alzheimer's Disease Education and Referral Center
http://www.alzheimers.org

American Academy of Neurology
1080 Montreal Avenue
St. Paul, MN 55116-2325
612-695-1940
http://www.aan.com
(For information on Alzheimer's disease, Parkinson's disease, migraine)

American College for the Advancement of Medicine
23121 Vergudo Drive, Suite 204
Laguna Hills, CA 92653
http://www.acam.org
1-949-583-7666
Physician referrals

American Heart Association
7272 Greenville Avenue
Dallas, TX 75231
Heart and stroke information at 1-800-AHA-USA1
http://www.americanheart.org

American Liver Foundation
75 Maiden Lane, Suite 603
New York, NY 10038
1-800-GOLIVER (1-800-465-4837)
http://www.liverfoundation.org

Arthritis Foundation
1330 W. Peachtree Street
Atlanta, GA 30309
1-800-283-7800
http://www.arthritis.org

Arthritis Society
393 University Avenue, Suite 1700
Toronto, ON Canada M5G 1E6
1-416-979-7228
http://www.arthritis.ca

Fibromyalgia Alliance of America
PO Box 21990
Columbus, OH 43221-0990
1-614-457-4222

Fibromyalgia Network
PO Box 31750
Tucson, AZ 85751-1750
1-800-853-2929

Hepatitis Center
http://www.hepatitis-central.com
A website for information on liver disease

National Foundation for Depressive Illness
PO Box 2257
New York, NY 10116
1-800-248-4344
http://www.depression.org

National Parkinson Foundation, Inc.
1501 NW 9th Avenue
Bob Hope Road
Miami, FL 33136-1494
1-800-327-4545

Parkinson's Action Network
822 College Avenue, Suite C
Santa Rosa, CA 95404
1-800-850-4726

Additional Reading

Arthritis Foundation. *Your Personal Guide to Living Well with Fibromyalgia*. Atlanta: Longstreet Press, 1997.

Grazi, Sol, and Marie Costa. *The European Arthritis and Depression Breakthrough! SAMe*. Rocklin, CA: Prima Publishing, 1999.

Mindell, Earl. *Earl Mindell's Herb Bible*. New York: Simon & Schuster, 1992.

———. *Earl Mindell's Vitamin Bible*. New York: Warner Books, 1991.

Murray, Michael T., N.D. *Encyclopedia of Natural Medicine*. Rocklin, CA: Prima Publishing, 1998.

———. *Encyclopedia of Nutritional Supplements*. Rocklin, CA: Prima Publishing, 1996.

———. *The Healing Power of Herbs*. 2nd ed. Rocklin, CA: Prima Publishing, 1995.

———. *Natural Alternatives to Prozac*. New York: William Morrow, 1996.

The PDR Family Guide to Nutrition and Health. Montvale, NJ: Medical Economics, 1998.